STEEL DETAILERS' MANUAL

STEEL DETAILERS' MANUAL

ALAN HAYWARD
CEng, FICE, FIStructE, MIHT

and

FRANK WEARE
CEng, MSc (Eng), DIC, DMS, FIStructE, MICE, MIHT, MBIM

Third Edition revised by
ANTHONY OAKHILL
BSc, CEng, MICE

WILEY-BLACKWELL
A John Wiley & Sons, Ltd., Publication

Blackwell Publishing was acquired by John Wiley & Sons in February 2007. Blackwell's publishing programme has been merged with Wiley's global Scientific, Technical, and Medical business to form Wiley-Blackwell.

First edition published in hardback in 1989
Reprinted in paperback 1992
Second edition published 2002

Registered office
John Wiley & Sons Ltd, The Atrium, Southern Gate, Chichester, West Sussex, PO19 8SQ, United Kingdom

Editorial office
9600 Garsington Road, Oxford, OX4 2DQ, United Kingdom
The Atrium, Southern Gate, Chichester, West Sussex, PO19 8SQ, UK
2121 State Avenue, Ames, Iowa 50014-8300, USA

For details of our global editorial offices, for customer services and for information about how to apply for permission to reuse the copyright material in this book please see our website at www.wiley.com/wiley-blackwell.

Diagrams and details presented in this manual were prepared by structural draughtsmen employed in the offices of Cass Hayward and Partners, Consulting Engineers of Chepstow (Monmouthshire, UK), who are regularly employed in the detailing of structural steelwork for a variety of clients including public utilities, major design: build contractors and structural steel fabricators

Library of Congress Cataloging-in-Publication Data

Hayward, Alan /0 06382208
    Steel detailers' manual / ALAN HAYWARD CEng, FICE, FIStructE, MIHT and FRANK WEARE. – Third
Edition / revised by ANTHONY OAKHILL.
        pages cm
    Includes bibliographical references and index.
    ISBN 978-1-4051-7521-0 (hardcover : alk. paper) 1. Steel, Structural–Handbooks, manuals, etc.
2. Building, Iron and steel–Handbooks, manuals, etc. 3. Building, Iron and steel–Details–Handbooks,
manuals, etc. I. Weare, Frank. II. Oakhill, Anthony, C. Eng. III. Title.
    TA685.H39 2011
    624.1'821–dc22   2010042177

A catalogue record for this book is available from the British Library.

This book is published in the following electronic formats: ePDF [ISBN 9781444393262]; Wiley Online Library [ISBN 9781444393286]; ePub [ISBN 9781444393279].

Set in 9/14pt Trebuchet by Thomson Digital, Noida, India
Printed and bound in Malaysia by Vivar Printing Sdn Bhd

1   2011

Contents

List of Figures		vii
List of Tables		ix
Preface		xi

1 Use of Structural Steel — 1
- 1.1 Why steel? — 1
- 1.2 Structural steels — 2
  - 1.2.1 Requirements — 2
  - 1.2.2 Recommended grades — 3
  - 1.2.3 Weather resistant steels — 4
- 1.3 Structural shapes — 6
- 1.4 Tolerances — 8
  - 1.4.1 General — 8
  - 1.4.2 Worked examples – welding distortion for plate girder — 12
- 1.5 Connections — 15
- 1.6 Interface to foundations — 16
- 1.7 Welding — 16
  - 1.7.1 Weld types — 16
  - 1.7.2 Processes — 17
  - 1.7.3 Weld size — 18
  - 1.7.4 Choice of weld type — 19
  - 1.7.5 Lamellar tearing — 19
- 1.8 Bolting — 20
  - 1.8.1 General — 20
  - 1.8.2 High strength friction grip (HSFG) bolts — 22
  - 1.8.3 Tension control bolts — 22
  - 1.8.4 European bolting standards — 23
- 1.9 Dos and don'ts — 24
- 1.10 Protective treatment — 24
- 1.11 Drawings — 30
  - 1.11.1 Engineer's drawings — 30
  - 1.11.2 Workshop drawings — 33
  - 1.11.3 Computer aided detailing — 33
- 1.12 Codes of practice — 33
  - 1.12.1 Buildings — 34
  - 1.12.2 Bridges — 34

2 Detailing Practice — 36
- 2.1 General — 36
- 2.2 Layout of drawings — 36
- 2.3 Lettering — 36
- 2.4 Dimensions — 36
- 2.5 Projection — 37
- 2.6 Scales — 37
- 2.7 Revisions — 37
- 2.8 Beam and column detailing conventions — 37
- 2.9 Erection marks — 38
- 2.10 Opposite handing — 39
- 2.11 Welds — 39
- 2.12 Bolts — 39
- 2.13 Holding down bolts — 39
- 2.14 Abbreviations — 40

3 Design Guidance — 41
- 3.1 General — 41
- 3.2 Load capacities of simple connections — 41
- 3.3 Sizes and load capacity of simple column bases — 42

4 Detailing Data — 52

5 Connection Details — 84

6 Computer Aided Detailing — 95
- 6.1 Introduction — 95
- 6.2 Steelwork detailing — 95
- 6.3 Constructing a 3-D model of a steel structure — 97
- 6.4 Object orientation — 99
- 6.5 CNC/rapid prototyping — 99
- 6.6 Future developments — 101

7 Examples of Structures — 102
- 7.1 Multi-storey frame buildings — 102
  - 7.1.1 Fire resistance — 105
- 7.2 Single-storey frame buildings — 106

7.3	Portal frame buildings	107	**Table of Standards** **141**
7.4	Vessel support structure	110	
7.5	Roof over reservoir	114	**References** **143**
7.6	Tower	117	
7.7	Bridges	121	**Further Reading** **144**
7.8	Single-span highway bridge	128	
7.9	Highway sign gantry	135	**Appendix** **146**
7.10	Staircase	139	
			Index **167**

List of Figures

1.1	Principles of composite construction	2
1.2	Stress : strain curves for structural steels	3
1.3	Corrosion rates of unpainted steel	5
1.4	Rolled section sizes	6
1.5	Twisting of angles and channels	8
1.6	Structural shapes	9–10
1.7	Welding distortion	11
1.8	Tolerances	13
1.9	Welding distortion – worked example	14
1.10	Flange cusping	14
1.11	Extra fabrication precamber	14
1.12	Bottom flange site weld	14
1.13	Web site weld	14
1.14	Functions of connections	15
1.15	Typical moment : rotation behaviour of beam/column connections	15
1.16	Continuous and simple connections	16
1.17	Locations of site connections	16
1.18	Connections in hot rolled and hollow sections	17
1.19	Connections to foundations	17
1.20	Butt welds showing double V preparations	18
1.21	Fillet welds	18
1.22	Sequence of fabrication	18
1.23	Welding using lapped joints	19
1.24	Lamellar tearing	20
1.25	Black bolts and HSFG bolts	21
1.26	Use of 'Coronet' load indicator	22
1.27	Dos and don'ts	25
1.28	Dos and don'ts	26–27
1.29	Dos and don'ts – corrosion	28–29
2.1	Drawing sheets and marking system	37
2.2	Dimensioning and conventions	38
3.1	Simple connections	44
3.2	Simple column bases	48
4.1	Stairs, ladders and walkways	75–76
4.2	Highway and railway clearances	77
4.3	Maximum transport sizes	78
4.4	Weld symbols	79
4.5	Typical weld preparations	80–81
5.1	Typical beam/column connections	84
5.2	Typical beam/beam connections	85
5.3	Typical column top and splice detail	86
5.4	Typical beam splices and column bases	87
5.5	Typical bracing details	88
5.6	Typical hollow section connections	89
5.7	Typical truss details	90
5.8	Workshop drawing of lattice girder – 1	91
5.9	Workshop drawing of lattice girder – 2	92
5.10	Typical steel/timber connections	93
5.11	Typical steel/precast concrete connections	94
6.1	The central role of the 3-D modelling system	97
6.2	Typical standard steelwork connection library	98
6.3	CNC/rapid prototyping guide	100
7.1–7.6	Multi-storey frame buildings	103–105
7.7–7.8	Single-storey frame buildings	106–107
7.9–7.10	Portal frame buildings	108–110
7.11–7.14	Vessel support structure	111–114
7.15–7.16	Roof over reservoir	115–116
7.17–7.19	Tower	117–120
7.20–7.24	Bridges	122–127
7.25–7.30	Single-span highway bridge	129–134
7.31–7.33	Highway sign gantry	136–138
7.34–7.35	Staircase	139–140

List of Tables

1.1	Advantages of structural steel	1
1.2	Steels to EN material standards – summary of leading properties	4
1.3	Main use of steel grades	4
1.4	Guidance on steel grades in BS 5950 – 1:2000 – design strengths	5
1.5	Guidance on steel grades in BS 5950 – 1:2000 – maximum thicknesses	5
1.6	Comparison of new and old section designation systems	7
1.7	Sections curved about major axis – typical radii	7
1.8	Dimensional variations and detailing practice	12
1.9	Common weld processes	19
1.10	Bolts used in UK	21
1.11	Typical cost proportion of steel structures	24
1.12	National standards for grit blasting	30
1.13a	Typical protective treatment systems for building structures	31
1.13b	Summary table of Highways Agency painting specifications for Highway Works Series 1900 – 8th edition (1998)	32
1.14	BS 5950 Load factors γ f and combinations	34
2.1	Drawing sheet sizes	36
2.2	List of abbreviations	40
3.1	Simple connections, bolts grade 4.6, members grade S275	43
3.2	Simple connections, bolts grade 8.8, members grade S275	45
3.3	Simple connections, bolts grade 8.8, members grade S355	46
3.4	Simple column bases	47
3.5	Black bolt capacities	49
3.6	HSFG bolt capacities	50
3.7	Weld capacities	51
4.1	Dimensions of black bolts	53
4.2	Dimensions of HSFG bolts	54
4.3	Universal beams – to BS 4-1: 2005	55–57
4.4	Universal columns	58
4.5	Joists	59
4.6	Channels	60
4.7	Rolled steel angles	61–62
4.8	Square hollow sections	64
4.9	Rectangular hollow sections	65
4.10	Circular hollow sections	66-67
4.11	Metric bulb flats	68-69
4.12	Crane rails	69–70
4.13	Face clearances pitch and edge distance for bolts	72
4.14	Durbar floor plate	73–74
4.15	Plates supplied in the 'Normalised' condition	82
4.16	Plates supplied in the 'Normalised Rolled' and 'Thermo-Mechanically Controlled Rolled' condition	83

Preface

It is now almost 25 years since this manual was first published. Its purpose now, as then, is to provide an introduction and guide to those in the constructional steelwork industry who are likely to be involved with the principles concerning the detailing of structural steel. The third edition of this important detailing manual recognises the principal changes which have occurred over this period of time.

There continues to be a marked improvement in steel's market share for buildings and bridges, both here in the UK and in many overseas' countries. Design and construction engineers, and architects, have continued to develop their appreciation for the often striking and awe-inspiring structures that have been designed and built in steel. But for the general public, who first see and marvel at these buildings and bridges, the creation, planning and development of any new steel structure is largely an unknown story. The many hours of work required to transform a sketch, resulting from a brain-storming meeting, into shaped pieces of elegant steelwork, are for the most part not well understood or even appreciated by the public at large. But also what is less well understood is that the nature of steel construction has markedly changed. During that period, the mix has moved from being predominantly industrial to being predominantly commercial. Steelwork has most convincingly established itself as the modern day material, being without equal for the many highly-visible prestigious and stimulating structures which adorn our landscape throughout the country.

It is often said that simple sketches and drawings can often account for a multitude of words and, of course, it is the production of those drawings, the detailing of the steelwork structure that provides the unbroken link between the designer and the constructor. One of the most important functions of the detailed drawing is to demonstrate the anticipated costs of the proposed steelwork structure. The costs of steelwork are not just confined to the raw materials and the production of the basic steel sections, but are determined more importantly by the connection details. Steelwork contractors will often confirm that their businesses depend on economic detailing. It is here then that one of the most important roles in steelwork production rests in the control of the steelwork detailer or CAD technician.

Steelwork designers have had to come to terms with the advent and increasing use of European design and construction standards. The manual attempts to clarify the present situation. It is however recognised that this is a constantly changing target, and the reader is advised to consult British Standards and other recognised professional steelwork organisations to determine the latest information.

For the steelwork detailer perhaps the most important development in recent times has been the rise of 3-D modelling techniques, the increased use of drawing layers, and the ability to speedily transmit drawings electronically between offices, works and sites. By these methods, it means that all parties to a project can inspect and comment on the developing details with a minimum of delay, which helps with keeping costs in check.

Steelwork contractors have also become highly used to operating sophisticated numerically-controlled machinery to cut, saw, drill and weld plates and sections with a high degree of precision. Again it is the detailer who provides the required link between the aspirations of the designer, and the commercial objectives of the constructor.

The authors continue to acknowledge the advice and help given to them in the preparation of this manual by their many friends and colleagues in construction. In particular thanks are due to the Corus Construction Centre (now a subsidiary of Tata Steel Europe), the British Constructional Steelwork Association and the Steel Construction Institute who gave permission for use of data.

Anthony Oakhill

1 Use of Structural Steel

1.1 Why steel?

Structural steel has distinct capabilities compared with other construction materials such as reinforced concrete, prestressed concrete, timber and brickwork. In most structures it is used in combination with other materials, the attributes of each combining to form the whole. For example, a factory building will usually be steel framed with foundations, ground and suspended floors of reinforced concrete. Wall cladding might be of brickwork with the roof clad with profiled steel sheeting. Stability of the whole building usually relies upon the steel frame, some-times aided by inherent stiffness of floors and cladding. The structural design and detailing of the building must consider this carefully and take into account intended sequences of construction and erection. Compared with other media, structural steel has attributes as given in Table 1.1.

In many projects the steel frame can be fabricated while the site construction of foundations is being carried out. Steel is also very suitable for phased construction which is a necessity on complex projects. This will often lead to a shorter construction period and an earlier completion date.

Table 1.1 Advantages of structural steel.

Feature	Leading to	Advantage	
		in buildings	in bridges
1. Speed of construction	Quick erection to full height of self supporting skeleton	Can be occupied sooner	Less disruption to public
2. Adaptability	Future extension	Flexible planning for future	Ability to upgrade for heavier loads
3. Low construction depth	Reduced height of structure	Cheaper heating Reduced environmental effect	Cheaper earthworks Slender appearance
4. Long spans	Fewer columns	Flexible occupancy	Cheaper foundations
5. Permanent slab formwork	Falsework eliminated	Finishes start sooner	Less disruption to public
6. Low weight of structure	Fewer piles and size of foundations Typical 50% weight reduction over concrete	Cheaper foundations and site costs	
7. Prefabrication in workshop	Quality control in good conditions, avoiding sites affected by weather	More reliable product Fewer specialist site operatives needed	
8. Predictable maintenance costs	Commuted maintenance costs can be calculated. If repainting is made easy by good design, no other maintenance is necessary	Total life cost known Choice of colour	
9. Lightweight units for erection	Erection by smaller cranes	Reduced site costs	
10. Options for site joint locations	Easy to form assemblies from small components taken to remote sites	Flexible construction planning	

Steel Detailers' Manual, Third Edition. Alan Hayward and Frank Weare. Third edition revised by Anthony Oakhill.
© 2011 Alan Hayward, Frank Weare and Anthony Oakhill. Published 2011 by Blackwell Publishing Ltd.

Steel is the most versatile of the traditional construction materials and the most reliable in terms of consistent quality. By its very nature it is also the strongest and may be used to span long distances with a relatively low self weight. Using modern techniques for corrosion protection the use of steel provides structures having a long reliable life, and allied with use of fewer internal columns achieves flexibility for future occupancies. Eventually when the useful life of the structure is over, the steelwork may be dismantled and realise a significant residual value not achieved with alternative materials. There are also many cases where steel frames have been used again, re-erected elsewhere.

Structural steel can, in the form of composite construction, co-operate with concrete to form members which exploit the advantages of both materials. The most frequent application is building floors or bridge decks where steel beams support and act compositely with a concrete slab via shear connectors attached to the top flange. The compressive capability of concrete is exploited to act as part of the beam upper flange, tension being resisted by the lower steel flange and web. This results in smaller deflections than those to be expected for non-composite members of similar cross-sectional dimensions. Economy results because of best use of the two materials – concrete which is effective in compression – and steel which is fully efficient when under tension. The principles of composite construction for beams are illustrated in figure 1.1 where the concept of stacked plates shown in (a) and (b) illustrates that much greater deflections occur when the plates are horizontal and slip between them can occur due to bending action. In composite construction relative slip is prevented by shear connectors which resist the horizontal shear created and which prevent any tendency of the slab to lift off the beam.

Structural steel is a material having very wide capabilities and is compatible with and can be joined to most other materials, including plain concrete, reinforced or pre-stressed concrete, brickwork, timber, plastics and earthenware. Its co-efficient of thermal expansion is virtually identical with that of concrete so that differential movements from changes in temperature are not a serious consideration when these materials are combined. Steel is often in competition with other materials, particularly structural concrete. For some projects different contractors often compete to build the structural frame in steel or concrete to maximise use of their own particular skills and resources. This is healthy as a means of maintaining reasonable construction costs. Steel though is able to contribute effectively in almost any structural project to a significant extent.

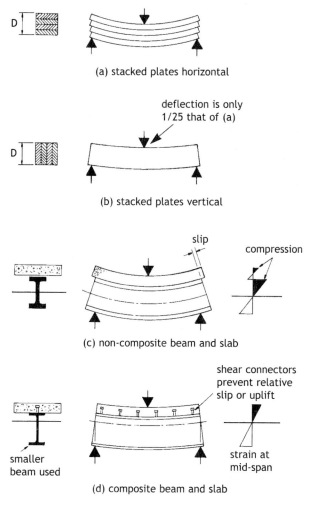

Figure 1.1   Principles of composite construction.

1.2   Structural steels

1.2.1   Requirements

Steel for structural use is normally hot rolled from billets in the form of flat plate or section at a rolling mill by the steel producer, and then delivered to a steel fabricator's workshop, where components are manufactured to precise form with connections for joining them together at site. Frequently used sizes and grades are also supplied by the mills to steel stockholders from whom fabricators may conveniently purchase material at short notice, but often at higher cost. Fabrication involves operations of sawing, shearing, punching, grinding, bending, drilling and welding to the steel so that it must be suitable for undergoing these processes without detriment to its required properties. It must possess reliable and predictable strength so that structures may be safely designed to carry the specified loads. The cost : strength ratio must be as low as possible consistent with these requirements to achieve economy. Structural steel must possess sufficient ductility so as to give warning (by visible deflection) before collapse conditions are reached in any structure which becomes unintentionally loaded beyond its design capacity and to allow use of fabrication processes such as cold bending. The ductility of structural steel is a particular attribute which is exploited where the 'plastic' design method is used for continuous (or statically indeterminate) structures in which significant deformation of the structure is implicit at factored loading. Provided that restraint against buckling is ensured this enables a structure to carry greater predicted loadings compared with the 'elastic' approach (which limits the maximum capacity to when yield stress is first reached at the most highly stressed fibre). The greater capacity is achieved by redistribution of forces and stress in a continuous structure, and by the contribution of the entire cross section at yield stress to resist the applied bending. Ductility may be defined as the ability of the material to elongate (or strain) when stressed beyond its yield limit, shown as the strain plateau in figure 1.2. Two measures of ductility are the 'elongation' (or total strain at fracture) and the ratio of ultimate strength to yield strength. For structural steels these values should be at least 18 per cent and 1.2 respectively.

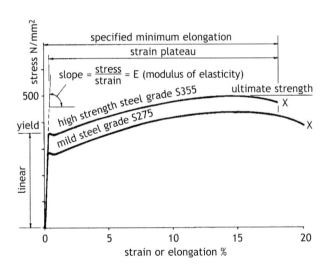

Figure 1.2 Stress : strain curves for structural steels.

For external structures in cold environments (i.e. typically in countries where temperatures less than about 0 °C are experienced) then the phenomenon of *brittle fracture* must be guarded against. Brittle fracture will only occur if the following three situations are realised simultaneously:

(1) a high tensile stress.
(2) low temperature.
(3) A notch-like defect or other 'stress raiser' exists.

The stress raiser can be caused by an abrupt change in cross section, a weld discontinuity, or a rolled-in defect within the steel. Brittle fracture can be overcome by specifying a steel with known 'notch ductility' properties, usually identified by the 'Charpy V-notch' impact test, measured in terms of energy in joules at the minimum temperature specified for the project location.

These requirements mean that structural steels need to be weldable low carbon type. In many countries a choice of mild steel or high strength steel grades are available with comparable properties. In the UK as in the rest of Europe structural steel is now obtained to EN 10025 (which, with other steel Euronorm standards, has replaced British Standards). Mild steel grades, previously 43A, 43B, etc., are now designated S275. High tensile steel grades, previously 50A, 50B, etc. are now referred to as S355. The grades are further designated by a series of letters (e.g. S275JR, S355JO) which denotes the requirements for Charpy V-notch impact testing. There is no requirement for impact testing for those grades which contain no letter. For other grades a different set of letters denotes an increased requirement (i.e. tested at a lower temperature). The main properties for the most commonly used grades are summarised in Table 1.2.

1.2.2 Recommended grades

In general it is economic to use high strength steel grade S355 due to its favourable cost : strength ratio compared with mild steel grade S275 typically showing a 20% advantage. Where deflection limitations dictate a larger member size (such as in crane girders) then it is more economic to use mild steel grade S275 which is also convenient for very small projects or where the weight in a particular size is less than, say 5 tonnes, giving choice in obtaining material from a stockholder at short notice.

Accepted practice is to substitute a higher grade in case of non-availability of a particular steel, but in such cases it is important to show the actual grade used on workshop drawings because different weld procedures may be necessary. Grades S420 and S460 offer a higher yield strength than grade S355, but they have not been widely used except for crane jibs and large bridge structures. Table 1.3 shows typical use of steel grades and guidance is given in Tables 1.4 and 1.5, the requirements for maximum thickness being based upon BS 5950 for buildings. BS 5400 for bridges has similar requirements.

Table 1.2 Steels to EN material standards – summary of leading properties.

| Grade | Tensile strength N/mm^2 | Yield strength N/mm^2 | Impact energy (J°C) | | |
| | | | Nominal thickness | | |
			Temp °C	≤150 mm	>150 ≤150 mm
S235	340/470	235	–	–	–
S235JR (1)	340/470 (1)	235	+20	27	–
S235JRG1 (1)	340/470 (1)	235	+20	27	–
S235JRG2	340/470	235	+20	27	23
S235JO	340/470	235	0	27	23
S235J2G3	340/470	235	−20	27	23
S235J2G4	340/470	235	−20	27	23
S275	410/560	275	–	–	–
S275	410/560	275	+20	27	23
S275	410/560	275	0	27	23
S275	410/560	275	−20	27	23
S275	410/560	275	−20	27	23
S355	490/630	355	–	–	–
S355JR	490/630	355	+20	27	23
S355JO	490/630	355	0	27	23
S355J2G3	490/630	355	−20	27	23
S355J2G4	490/630	355	−20	27	23
S355K2G3	490/630	355	−20	40	33
S355K2G4	490/630	355	−20	40	33

(1) Only available up to and including 25 mm thick.

Other properties of steel:

Modulus of elasticity	$E = 205 \times 10^3$ N/mm^2 (205 kg/mm^2)
Coefficient of thermal expansion	12×10^6 per °C
Density or mass	7850 kg/m^3 (7.85 tonnes/m^3 or 78.5 kN/m^3)
Elongation (200 mm gauge length)	

Grade S275	20%
S355	18%
S460	17%
S355JOW	19%

Table 1.3 Main use of steel grades.

	BS EN 10025 BS EN 10113 (Pts 1 & 2)	Yield N/mm^2	As rolled cost : strength ratio	Type
Buildings	S275	275	1.00	Mild
	S355	355	0.84	High Strength
Bridges	S420	420	–	Ditto
	S460	460	0.81	Ditto
Cranes	S690 (BS EN 10137)	690	–	Ditto

1.2.3 Weather resistant steels

When exposed to the atmosphere, low carbon equivalent structural steels corrode by oxidation forming rust and this process will continue and eventually reduce the effective thickness leading to loss of capacity or failure. Stainless steels containing high percentages of alloying elements such as chromium and nickel can be used to minimise the corrosion process but their very high cost is virtually prohibitive for most structural purposes, except for small items such as bolts in critical locations. Protective treatment systems are generally applied to structural steel frameworks using a combination of painting, metal spraying or galvanising, depending upon the environmental conditions and ease of future maintenance. Costs of maintenance can be significant for structures having difficult access conditions, such as high-rise buildings with exposed frames and for bridges.

Table 1.4 Guidance on steel grades in BS 5950 – 1 : 2000 – design strengths.

Steel grade	Thickness* less than or equal to mm	Design strength p_y N/mm^2
S275	16	275
	40	265
	63	255
	80	245
	100	235
	150	225
S355	16	355
	40	345
	63	335
	80	325
	100	315
	150	295
S460	16	460
	40	440
	63	430
	80	410
	100	400

*For rolled sections, use the specified thickness of the thickest element of the cross-section.

Weather resistant steels which develop their own corrosion resistance and which do not require protective treatment or maintenance were developed for this reason. They were first used for the John Deare Building in Illinois in 1961, the exterior of which consists entirely of exposed steelwork and glass panels; several prestigious buildings have since used weather resistant steel frames. The first bridge was built in 1964 in Detroit followed by many more in North America and several hundred UK bridges have been completed since 1968. Costs of weather resistant steel frames tend to be marginally greater due to a higher material cost per tonne, but this may more than offset the alternative costs of providing protective treatment and its long term maintenance. Thus weather resistant steel deserves consideration where access for maintenance will be difficult.

Weather resistant steels contain up to 3 per cent of alloying elements such as copper, chromium, vanadium and phosphorous. The steel oxidises naturally and when a tight patina of rust has formed this inhibits further corrosion. Figure 1.3 shows relative rates of corrosion. Over a period of one to four years the steel weathers to a shade of dark brown or purple depending upon the atmospheric conditions in the locality. Appearance is enhanced if the steel has been blast cleaned after fabrication so that weathering occurs evenly.

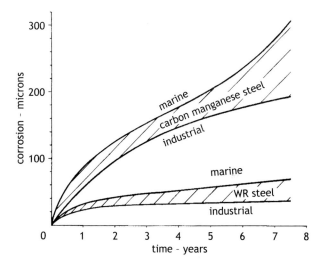

Figure 1.3 Corrosion rates of unpainted steel.

BS EN 10155 gives the specific requirements for the chemical composition and mechanical properties of the S355JOW grades rolled in the UK, which are similar to Corten B as originated in the USA. Because the material is less widely used weather resistant steels are not widely available from stockholders. Therefore small tonnages for a particular rolled section should be avoided. There are a few stockholders who will supply a limited range of rolled plate. Welding procedures need to be more stringent than for

Table 1.5 Guidance on steel grades in BS 5950 – 1:2000 – maximum thicknesses*.

Product standard	Steel grade or quality	Sections	Plates and flats	Hollow sections
BS EN 10025	S275 *or* S355	100	150	–
BS EN 10113-2	S275 *or* S355	150	150	–
	S460	100	100	–
BS EN 10113-3	S275, S355 *or* S460	150	63	–
BS EN 10137-2	S460	–	150	–
BS EN 10155	J0WP *or* J2WP	40	16	–
	J0W, J2W *or* K2W	100	100	–
BS EN 10210-1	All	–	–	65
BS EN 10219-1	All	–	–	40
BS 7668	J0WPH	–	–	12
	J0WH *or* GWH	–	–	40

*Maxi. thickness at which the full Charpy impact value given in the product standard applies.

other high tensile steel due to the higher carbon equivalent, and it must be ensured that exposed weld metal has equivalent weathering properties. Suitable alloy-bearing consumables are available for common welding processes, but for single run welds using manual or submerged arc it has been shown that sufficient dilution normally occurs such that normal electrodes are satisfactory. It is only necessary for the capping runs of butt welds to use electrodes with weathering properties.

Until the corrosion inhibiting patina has formed it should be realised that rusting takes place and run-off will occur, which may cause staining of concrete and other parts locally. This can be minimised by careful attention to detail. A suitable drip detail for a bridge is shown in figure 7.27. Drainage of pier tops should be provided to prevent streaking of concrete and, during construction, temporary protection specified. Weather resistant steels are not suitable in conditions of total immersion or burial and therefore water traps should be avoided and columns terminated above ground level. Use of concrete or other light coloured paving should be avoided around column bases, and dark coloured brickwork or gravel finish should be considered. In the UK it is usual in bridges to design[1] against possible long term slow rusting of the steel by added thicknesses (1.5 mm for exposed face in very severe environments and 1 mm otherwise), severity being a function of the atmospheric sulphur level. Weather resistant steel should not be used in marine environments and water containing chlorides such as de-icing salts should be prevented from coming into contact by suitable detailing. At expansion joints on bridges consideration should be given to casting in concrete locally in case of leakage as shown in figure 7.27.

Extra care must be taken in materials ordering and control during the fabrication of projects in weathering steel because its visual appearance is similar to other steels during manufacture. Testing methods are available for identification of material which may have been inadvertently misplaced.

1.3 Structural shapes

Most structures utilise hot rolled sections in the form of universal beams (UBs), universal columns (UCs), channels and rolled steel angles (RSAs) to BS 4, see figure 1.6. Less frequently used are tees cut from universal beams or columns such that the depth is one half of the original section. Hollow sections in the form of circular (CHS), square (SHS)

and rectangular (RHS) shape are available but their cost per tonne is approximately 20 per cent more than universal beams and columns. Although efficient as struts or columns, the end connections tend to be complex especially when bolted. They are often used where clean appearance is vital, such as steelwork which is exposed to view in public buildings. Wind resistance is less that of open sections giving an advantage in open braced structures such as towers, where the steelwork itself contributes to most of the exposed area. Other sections are available such as bulb flats and trapezoidal troughs as used in stiffened plate construction, for example box girder bridges and ships.

The range of UBs and UCs offers a number of section weights within each serial size (depth D and breadth B). Heavier sections are produced with the finishing rolls further apart such that the overall depth and breadth increase, but with the clear distance between flanges remaining constant, as shown in figure 1.4. This is convenient in multi-storey buildings in allowing use of lighter sections of the same serial size for the upper levels. However, it must be remembered that the actual overall dimensions (D and B) will often be greater than the serial size except when the basic (usually lightest) section is used. This will affect detailing and overall cladding dimensions. Drawings must therefore state actual dimensions. For other sections (e.g. angles and hollow sections) the overall dimensions (D and B) are constant for all weights within each serial size.

In 2006 Tata Steel Europe (formerly Corus Group) in the UK introduced its Advance section range to reflect the need for Corus CE-marked structural sections to comply with the requirements of the EU Directive on Construction Products. Twenty-one additional beams and columns have been added to the standard Corus UK section range to create the new Advance range. To simplify specification

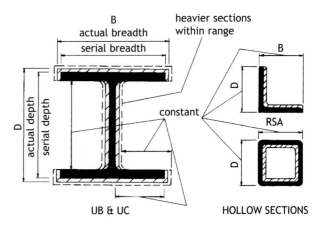

Figure 1.4 Rolled section sizes.

Table 1.6 Comparison of new and old section designation systems.

Corus Advance sections		Old designation system	
UKB	UK Beam	UB	Universal Beam
UKC	UK Column	UC	Universal Column
UKPFC	UK Parallel Flange Channel	PFC	Parallel Flange Channel
UKA	UK Angle	RSA	Rolled Steel Angle
UKBP	UK Bearing Pile	UBP	Universal Bearing Pile
UKT	UK Tee		

Example: $457 \times 191 \times 67$UB becomes $457 \times 191 \times 67$UKB

Table 1.7 Sections curved about major axis – typical radii.

	Typical radius (curved about major axis)	
Section size	Cold bending	Hot bending
$838 \times 292 \times 226$ UB	75000 mm	12500 mm
$762 \times 267 \times 197$ UB	50000 mm	10000 mm
$610 \times 305 \times 238$ UB	25000 mm	8000 mm
$533 \times 210 \times 82$ UB	25000 mm	5000 mm
$457 \times 191 \times 74$ UB	20000 mm	4500 mm
$356 \times 171 \times 67$ UB	10000 mm	3000 mm
$305 \times 305 \times 137$ UC	10000 mm	2500 mm
$254 \times 254 \times 89$ UC	6000 mm	2500 mm
$203 \times 203 \times 60$ UC	4000 mm	1750 mm
$152 \times 152 \times 37$ UC	2000 mm	1250 mm

Information in this table is supplied by The Angle Ring Co. Ltd, Bloomfield Road, Tipton, West Midlands DY4 9EH, UK. Email: technical@anglering.co.uk.

of Advance sections, a new UK prefix has been introduced (as shown in Table 1.6).

Other rolled sections are available in the UK and elsewhere, including rails (for travelling cranes and railway tracks), bearing piles (H pile or welded box) and sheet piles (Larssen or Frodingham interlocking). Cellform (or castellated) beams are made from universal beam or column sections cut to corrugated profile and reformed by welding to give a 50 per cent deeper section providing an efficient beam for light loading conditions.

Sections sometimes need to be curved about one or both axes to provide precamber (to counteract dead load deflection of long span beams) or to achieve permanent curvature, for example in arched roofs or circular cofferdams. Specialists in the UK can curve structural steel sections by either cold (roller bending) or hot (induction bending) processes. In general, they can be curved to single-radius curves, to multi-radius curves, to parabolic or elliptical curves or even to co-ordinates. They can also, within limits, be curved in two planes or to form spirals.

The curving process has merit in that most residual stresses (inherent in rolled sections when produced) are removed such that any subsequent heat-inducing operations such as welding or galvanizing cause less distortion than otherwise. Although, usually more costly than cold rolling, hot induction bending enables steel sections to be curved to a very much smaller radius and with much less deformation, as indicated in Table 1.7. The minimum radius to which any section can be curved depends on its metallurgical properties (particularly ductility), its thickness, its cross-sectional geometry and the bending method. Table 1.7 gives typical radii to which a range of common sections can readily be curved about their major axes by cold or hot bending. Note that these are not minimum values so guidance on the

realistic minimum radii with regard to specific sections should be sought from a specialist bending company.

Other general guidelines include:

- small sections can, logically, be curved to smaller radii than larger ones
- within any one serial size, the heavier sections can normally be curved to a smaller radius than the lighter section
- universal columns can be curved to smaller radii about the major axis than universal beams of the same depth but, generally, the reverse applies about the minor axis
- most open sections (angles, channels) can be curved to a smaller radius about the minor axis than about the major axis.

Fabricated members are used for spans or loads in excess of the capacity of rolled sections. Costs per tonne are higher because of the extra operations in profile cutting and welding. Box girders have particular application where their inherent torsional rigidity can be exploited, for example in a sharply curved bridge. Compound members made from two or more interconnected rolled sections can be convenient, such as twin universal beams. For sections which are asymmetric about their major (x–x) axis, such as channels or rolled steel angles (RSAs) then interconnection or torsional restraint is a necessity if used as a beam. This is to avoid torsional instability where the shear centre of the section does not coincide with the line of action of the applied load as shown in figure 1.5.

Cold formed sections using thin gauge material (1.5 mm to 3.2 mm thick typically) are used for lightly loaded secondary

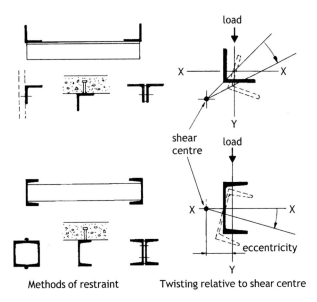

load

shear
centre

load

X ———— X

X ———— X

eccentricity

Y

Methods of restraint   Twisting relative to shear centre

Figure 1.5 Twisting of angles and channels.

members, such as purlins and sheeting rails. They are not suitable for external use. They are available from a number of manufacturers to dimensions particular to the supplier and are usually galvanised. Ranges of standard fitments such as sag rods, fixing cleats, cleader angles, gable posts and rafter stays are provided, such that for a typical single storey building only the primary members might be hot rolled sections. Detailing of cold rolled sections is not covered in this manual, but it is important that the designer ensures that stability is provided by these elements or if necessary provides additional restraint.

Open braced structures such as trusses, lattice or Vierendeel girders and towers or space frames are formed from individual members of either hot rolled, hollow, fabricated or compound shapes. They are appropriate for lightly loaded long span structures such as roofs or where wind resistance must be minimised, as in towers. In the past they were used for heavy applications such as bridges, but the advent of automated fabrication together with availability of wide plates means that plate girders are more economic.

1.4   Tolerances

1.4.1   General

In all areas of engineering the designer, detailer and constructor need to allow for tolerances. This is because in practice absolute precision cannot be guaranteed for each and every dimension even when working to very high

manufacturing standards. Very close tolerances are demanded in mechanical engineering applications where moving parts are involved and the high costs involved in machining operations after manufacture of such components have to be justified. Even here tolerance allowances are necessary and it is common practice for values to be specified on drawings. In structural steelwork such close tolerances could only be obtained at very high cost, taking into account the large size of many components and the variations normally obtained with rolled steel products. Therefore accepted practice in the interests of economy is to fabricate steelwork to reasonable standards obtainable in average workshop conditions and to detail joints which can absorb small variations at site. Where justified, operations such as machining of member ends after fabrication to precise length and/or angularity are carried out, but this is exceptional and can only be carried out by specialist fabricators. Normally, machining operations should be restricted to small components (such as tapered bearing plates) which can be carried out by a specialist machine shop remote from the main workshop and attached before delivery to site.

Many workshops have installed numerically controlled (NC) equipment for marking, sawing members to length, for hole drilling and profile cutting of plates to shape. This has largely replaced the need to make wooden (or other) templates to ensure fit-up between adjacent connections when preparation (i.e. marking, cutting and drilling) was performed by manual methods. Use of NC equipment has significantly improved accuracy such that better tolerances are achieved without need for adjustments by dressing or reaming of holes. However, the main factor causing dimensional variation is *welding distortion*, which arises due to shrinkage of the molten weld metal when cooling. The amount of distortion which occurs is a function of the weld size, heat input of the process, number of runs, the degree of restraint present and the material thicknesses.

To an extent *welding distortion* can be predicted and the effects allowed for in advance, but some fabricators prefer to exclude the use of welding for beam/column structures and to use all bolted connections. However, welding is necessary for fabricated sections such that the effects of distortion must be understood and catered for.

Figure 1.7 illustrates various forms of welding distortion and how they should be allowed for either by presetting, using temporary restraints or initially preparing elements with extra length. This is often done at workshop floor level, and

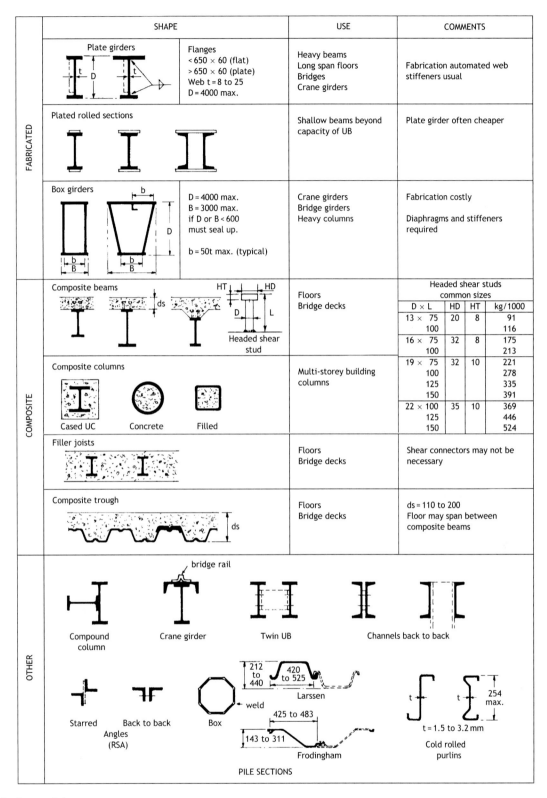

Figure 1.6 Structural shapes.

ideally should be calculated in consultation with the welding engineer and detailer. Where site welding is involved then the *workshop drawings* should include for weld shrinkage at site by detailing the components with extra length. A worked example is given in 1.4.2.

When site welding plate girder splices the flanges should be welded first so that shrinkage of the joint occurs before the (normally thinner) web joint is made, to avoid buckling. Therefore the web should be detailed with approximately 2 mm extra root gap, as shown in figure 1.13.

	SHAPE	UK SIZE RANGE	USE	COMMENTS
HOT ROLLED SECTIONS	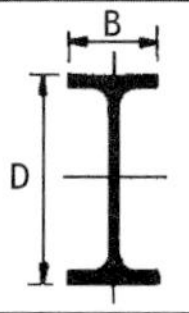 Universal beam (UB)	$D \times B \times$ kg/m — 127 × 76 × 13 to 914 × 419 × 388	Beams	May need bearing stiffeners at supports and under point loads — $B \times D$ = serial size. Actual dimensions vary with weight (kg/m)
	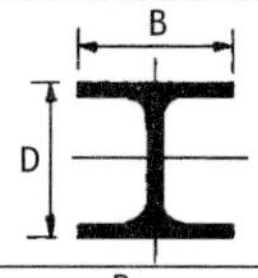 Universal column (UC) — Bearing pile or H-pile	152 × 152 × 23 to 356 × 406 × 634	Columns, Shallow beams, Heavy truss members, Piles	
	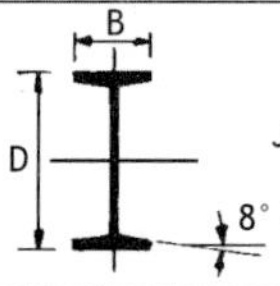 Joist — 8° taper	76 × 76 × 12.65 to 254 × 203 × 81.85	Small beams	
	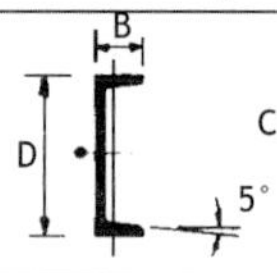 Channel — 5° taper	76 × 38 × 6.70 to 432 × 102 × 65.54	Bracings, Ties, Light beams	When used as beam torsion occurs relative to shear centre. Restrain or use in pairs e.g. 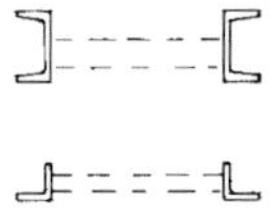
	 Equal angle (RSA) — shear centre	$D \times B \times t$ — 25 × 25 × 3 to 250 × 250 × 35	Bracings, Truss members, Tower members, Purlins, Sheeting rails	
	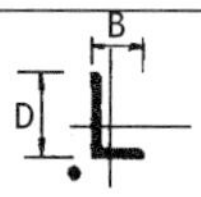 Unequal angle (RSA)	40 × 25 × 4 to 200 × 150 × 18		
	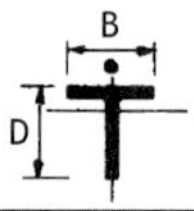 Structural tee	$D \times B \times$ kg/m — UB 76 × 64 × 7 to 419 × 457 × 194 — UC 152 × 76 × 12 to 406 × 178 × 317	Truss chords, Plate stiffeners	Cut from UB or UC. $D = 0.5 \times$ original depth
	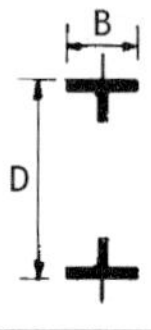 Castellated beam 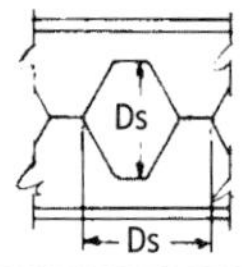	191 × 76 × 13 to 1371 × 419 × 388	Light beams where services need to pass through beam	Made from UB. $D = 1.5 \times D_s$ (approx)
	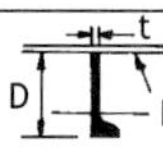 Bulb flat — plate	120 × 6 × 7.31 to 430 × 20 × 90.8	Plate stiffeners in bridges or pontoons etc.	Good welding access
HOLLOW SECTIONS	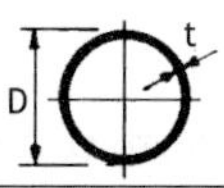 Circular hollow sections (CHS) and tubes	$D \times t$ — CHS 21.3 × 3.2 to 508 × 50 tubes up to 2020 × 25	Space frames, Columns, Bracings, Piles	End connections costly if bolted splice. Supply cost per tonne approx 20% higher than open sections
	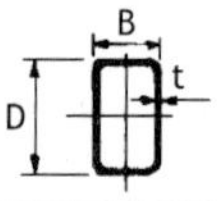 Rectangular hollow section (RHS)	$D \times B \times t$ — 50 × 25 × 2.5 to 500 × 300 × 20	Columns, Vierendeel girders	Fully seal ends if external use
	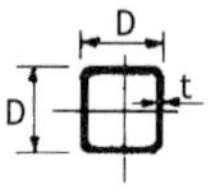 Square hollow section (SHS)	20 × 20 × 2 to 400 × 400 × 20	Columns, Space frames	Clean appearance

Figure 1.6   *Contd*

Table 1.8 shows some of the main causes of dimensional variations which can occur and how they should be overcome in detailing. These practices are well accepted by designers, detailers and fabricators. It is not usual to incorporate tolerance limits on detailed drawings although this will be justified in special circumstances where accuracy is vital to connected mechanical equipment. Figure 1.8 shows tolerances for rolled sections and fabricated members.

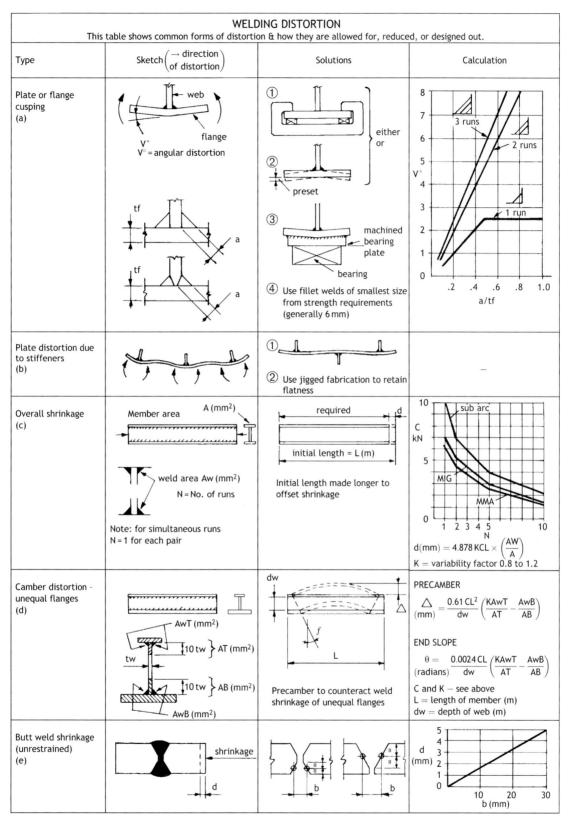

Figure 1.7 Welding distortion.

Table 1.8 Dimensional variations and detailing practice.

Type of variation	Detailing practice
1. Rolled sections – tolerances	Dimensions from top of beams down from centre of web Backmark of angles and channels
2. Length of members	Tolerance gap at ends of beams. Use lapped connections not abutting end plates For multi-storey frames with several bays consider variable tolerance packs
3. Bolted end connections	Black bolts or HSFG bolts in clearance holes For bolt groups use NC drilling or templates For large complex joints drill pilot holes and ream out to full size during a trial erection Provide large diameter holes and washer plates if excessive variation possible
4. Camber or straightness variation in members	Tolerance gap to beam splices nominal 6 mm Use lapped connections
5. Inaccuracy in setting foundations and holding down bolts to line and level	Provide grouted space below baseplates. Cast holding down bolts in pockets. Provide extra length bolts with excess thread
6. Countersunk bolts/ set screws	Avoid wherever possible
7. Weld size variation	Keep details clear in case welds are oversized
8. Columns prepared for end bearing	Machine ends of fabricated columns (end plates must be ordered extra thick) Incorporate division plate between column lengths
9. Cumulative effects on large structures	Where erection is costly or overseas delivery carry out trial erection of part or complete structure For closing piece on long structure such as bridge, fabricate or trim element to site measured dimensions
10. Fit of accurate mechanical parts to structural steelwork	Use separate bolted-on fabrication

1.4.2 Worked example – welding distortion for plate girder

Calculation of welding distortion

The following example illustrates use of figure 1.7 in making allowances for welding distortion for the welded plate girder shown in figure 7.28.

Worked example

Question

The plate girder has unequal flanges and is 32.55 m long over end plates. Web/flange welds are 8 mm fillet welds which should use the submerged arc process, each completed in a single run, but not concurrently on either side of web. For simplicity the plate sizes as at mid length are assumed to apply full length. The girder as simplified is shown in figure 1.9.

It is required to calculate:

(1) Amount of flange plate cusping which may occur due to web/flange welds.

(2) Additional length of plates to counteract overall shrinkage in length due to web/flange welds.

(3) Camber distortion due to unequal flanges so that extra fabrication precamber can be determined.

(4) Butt weld shrinkage for site welded splice so that girders can be detailed with extra length.

Answer

(1) *Amount of flange cusping*

Using figure 1.7(a)

Top flange: $a = 8 \div \sqrt{2} = 5.65$ mm weld throat
$tf = 25$ mm flange thickness
$\dfrac{a}{tf} = \dfrac{5.65}{25} = 0.226$ N = 1 for each weld

From figure 1.7(a) V = 1.1°

Bottom flange: $a = 8 \div \sqrt{2} = 5.65$ mm
$tf = 50$ mm flange thickness
$\dfrac{a}{tf} = \dfrac{5.65}{50} = 0.113$ N = 1

From figure 1.7(a) V = 0.5°
The resulting flange cusping is shown in figure 1.10.

Use of 'strongbacks' or presetting as shown in figure 1.7(a) may need to be considered during fabrication, because although the cusps are not detrimental structurally they may affect details especially at splices and at bearings.

(2) *Overall shrinkage*

Using figure 1.7(c):
shortening d = 4.878 kCL (Aw/A)
where   C = 5.0 kN for N = 4 weld runs
L = 32.55 m
$Aw = \left(\dfrac{8 \times 8}{2}\right) \times 4\,No = 128\,mm^2$
$A = (500 \times 25) + (600 \times 50) + (1300 \times 14)$
$= 60\,700\,mm^2$
k = 0.8 to 1.2

For
k = 0.8 $d = 4.878 \times 0.8 \times 5.0 \times 32.55 \times \dfrac{128}{60\,700} = 1.3$ mm,
or for k = 1.2   d = 2.0 mm

Therefore overall length of plates must be increased by 2 mm.

(3) *Camber distortion*

Using figure 1.7(d)

$Precamber = \Delta = \dfrac{0.61\,CL^2}{dw}\left(\dfrac{kAwt}{AT} - \dfrac{AwB}{AB}\right)$

ROLLED SECTIONS			TOLERANCES & EFFECTS IN DETAILING	
Tolerances			**DIMENSIONS**	**SKETCH**
Type		Value		

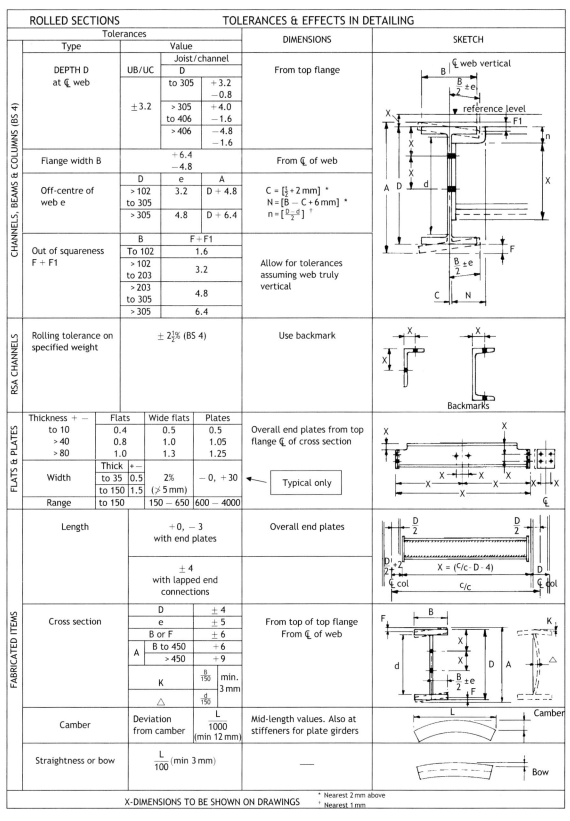

CHANNELS, BEAMS & COLUMNS (BS 4)

DEPTH D at ℄ web — Joist/channel

	UB/UC	D	
DEPTH D at ℄ web	±3.2	to 305	+3.2 / −0.8
		>305 to 406	+4.0 / −1.6
		>406	−4.8 / −1.6

From top flange

Flange width B: +6.4 / −4.8 — From ℄ of web

Off-centre of web e:

	D	e	A
	>102 to 305	3.2	D + 4.8
	>305	4.8	D + 6.4

$C = [\frac{t}{2} + 2\,mm]$ *
$N = [B - C + 6\,mm]$ *
$n = [\frac{D-d}{2}]$ †

Out of squareness F + F1:

B	F + F1
To 102	1.6
>102 to 203	3.2
>203 to 305	4.8
>305	6.4

Allow for tolerances assuming web truly vertical

RSA CHANNELS

Rolling tolerance on specified weight: ± 2½% (BS 4) — Use backmark

Backmarks

FLATS & PLATES

Thickness + −	Flats	Wide flats	Plates
to 10	0.4	0.5	0.5
>40	0.8	1.0	1.05
>80	1.0	1.3	1.25

Overall end plates from top flange ℄ of cross section

Width:

Thick	+ −		
to 35	0.5	2% (≯ 5 mm)	− 0, + 30
to 150	1.5		

Typical only

Range	to 150	150 − 650	600 − 4000

FABRICATED ITEMS

Length: +0, −3 with end plates — Overall end plates

±4 with lapped end connections

$X = (C/c - D - 4)$

Cross section:

D	±4
e	±5
B or F	±6
A B to 450	+6
A >450	+9
K	B/150 min. 3 mm
△	d/150

From top of top flange
From ℄ of web

Camber: Deviation from camber — L/1000 (min 12 mm) — Mid-length values. Also at stiffeners for plate girders

Straightness or bow: L/100 (min 3 mm) — — — Bow

X-DIMENSIONS TO BE SHOWN ON DRAWINGS

* Nearest 2 mm above
† Nearest 1 mm

Figure 1.8 Tolerances.

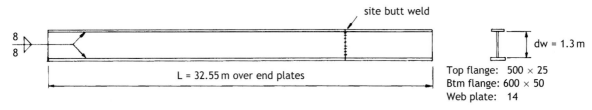

Figure 1.9 Welding distortion – worked example.

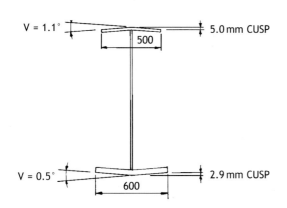

Figure 1.10 Flange cusping.

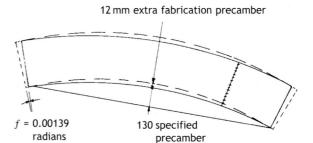

Figure 1.11 Extra fabrication precamber.

would be taken account of in materials ordering and during fabrication.

(4) Butt weld shrinkage

Using figure 1.7(e):
  bottom flange. See figure 1.12 for butt weld detail.
  shrinkage $d = 2.0$ mm

Therefore length of flanges must be increased by 1 mm on each side of splice and detailed as shown. Normal practice is to weld the flanges first. Thus the web will be welded under restraint and should be detailed with the root gap increased by 2 mm as shown in figure 1.13.

where C = 7.0 kN for N = 2 weld runs each flange
   L = 32.55 m
   dw = 1.30 m
   k = 0.8 to 1.2
   $AwT = \left(\dfrac{8 \times 8}{2}\right) \times 2 \text{ No} = 64 \text{ mm}^2$
   $AwB = 64 \text{ mm}^2$
   $AT = (500 \times 25) + (10 \times 14^2) = 14\,460 \text{ mm}^2$
   $AB = (600 \times 50) + (10 \times 14^2) = 31\,960 \text{ mm}^2$

For k = 0.8 $\Delta = \dfrac{0.61 \times 7.0 \times 32.55^2}{1.30}\left(\dfrac{0.8 \times 64}{14\,460} - \dfrac{64}{31\,960}\right)$
            $= 5.4$ mm

For k = 1.2 $\Delta = 11.5$ mm (say 12 mm)

End slope $\theta = \dfrac{0.0024\,CL}{dw}\left(\dfrac{kAwT}{AT} - \dfrac{AwB}{AB}\right)$

For k = 0.8 $\theta = \dfrac{0.0024 \times 7.0 \times 32.55}{1.30}\left(\dfrac{0.8 \times 64}{14\,460} - \dfrac{64}{31\,960}\right)$

            $= 0.00065$ radians

For k = 1.2 $\theta = 0.00139$ radians

Therefore extra fabrication precamber needs to be applied as shown in figure 1.11 additional to the total precamber specified for counteracting dead loads, etc. given in figure 7.25. This would not be shown on workshop drawings but

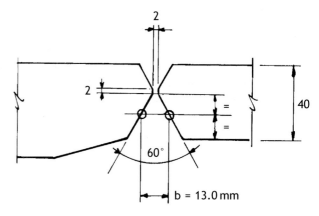

Figure 1.12 Bottom flange site weld.

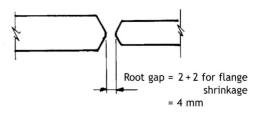

Figure 1.13 Web site weld.

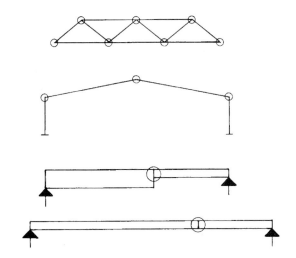

Figure 1.14 Functions of connections.

In producing workshop drawings in this case, only item (4) should be shown thereon because items (1) to (3) occur due to fabrication effects, which are allowed for at the workshop. Item (4) occurs at site and must therefore be taken into consideration so that the item delivered takes into account weld shrinkage at site.

1.5 Connections

Connections are required for the functions illustrated in figure 1.14. The number of site connections should be as few as possible consistent with maximum delivery/erection sizes so that the majority of assembly is performed under workshop conditions. Welded fabrication is usual in most workshops and is always used for members such as plate girders, box girders and stiffened platework.

It is always wise to consider the connection type to be used at the conceptual design stage. A *continuously* designed structure of lighter weight but with more complex fabrication work can be more expensive than a slightly heavier design with *simple* joints. Once the overall concept is decided the connections should always be given at least the same attention as the design of the main members which they form. Structural adequacy is not, in itself, the sole criterion because the designer must endeavour to provide an efficient and effective structure at the lowest cost.

With appropriate stiffening either an all welded or a high strength friction grip (HSFG) bolted connection is able to achieve a fully continuous joint, that is one which is capable of developing applied bending without significant rotation. However, such connections are costly to fabricate and erect. They may not always be justified. Many economical beam/column structures are built using angle cleat or welded end

plate connections without stiffening and then joined with *black bolts*. These are defined as simple connections which transmit shear but where moment/rotation stiffness is not sufficient to mobilise end fixity of beams or frame action under wind loading without significant deflection. Figure 1.15 shows typical moment : rotation behaviour of connections. Simple connections (i.e. types A or B) are significantly cheaper to fabricate although somewhat heavier beam sizes may be necessary because the benefits of end fixity leading to a smaller maximum bending moment are not realised. Use of simple connections enables the workshop to use automated methods more readily with greater facility for tolerance at site and will often give a more economic solution overall. However it is necessary to stabilise structures having simple connections against lateral loads such as wind by bracing or to rely on shear walls/lift cores, etc. For this reason simple connections should be made *erection-rigid* (i.e. retain resistance against free rotation whilst remaining flexible) so that the structure is stable during erection and before bracings or shear walls are connected. All connections shown in figure 1.15 are capable of

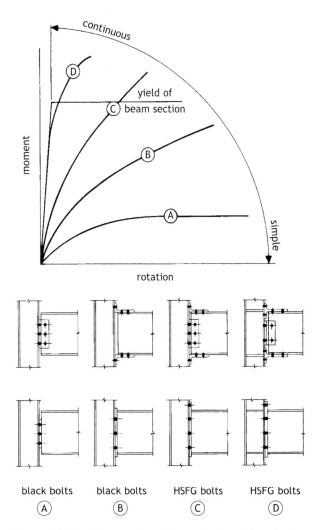

Figure 1.15 Typical moment : rotation behaviour of beam/column connections.

'continuous'

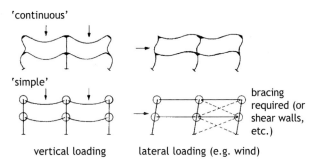

'simple'

bracing required (or shear walls, etc.)

vertical loading   lateral loading (e.g. wind)

Figure 1.16 Continuous and simple connections.

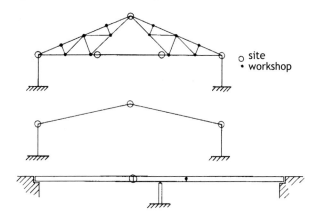

○ site
● workshop

Figure 1.17 Locations of site connections.

being erection-rigid. Calculations may be necessary in substantiation, but use of seating cleats only for beam/column connections should be avoided. A top flange cleat should be added. Web cleat or flexible (i.e. 12 mm maximum thickness) end plate connections of at least 0.6 × beam depth are suitable. Provision of seating cleats is not a theoretical necessity but they improve erection safety for high-rise structures exceeding 12 storeys. Behaviour of continuous and simple connections is shown in figure 1.16. Typical locations of site connections are shown in figure 1.17.

At site either welding or bolting is used, but the latter is faster and usually cheaper. Welding is more difficult on site because assemblies cannot be turned to permit downhand welding and erection costs arise for equipment in supporting/aligning connections, pre-heating/sheltering and non-destructive testing (NDT). The exception is a major project where such costs can be absorbed within a larger number of connections (say, minimum 500). As a general rule welding and bolting are used thus:

Welding – workshop

Bolting – site

For bridges continuous connections should be used to withstand vibration from vehicular loading and spans should usually be made continuous. This allows the numbers of

deck expansion joints and bearings to be reduced thus minimising maintenance of these costly items, which are vulnerable to traffic and external environment.

For UK buildings, connection design is usually carried out by the fabricator with the member sizes and end reactions being specified on the engineer's drawings. It is important that all design assumptions are advised to the fabricator for him to design and detail the connections. If joints are continuous then bending moments and any axial loads must be specified in addition to end reactions. For simple connections the engineer must specify how stability is to be achieved, both during construction and finally when in service.

Connections to hollow sections are generally more costly and often demand butt welding rather than fillet welds. Bolted connections in hollow sections require extended end plates or gussets and sealing plates because internal access is not feasible for bolt tightening whereas channels or rolled steel angles (RSAs) can be connected by simple lap joints. Figure 1.18 compares typical welded or bolted connections.

1.6 Interface to foundations

It is important to recognise whether the interface of steelwork to foundations must rely on a moment (or rigid) form of connection or not.

Figure 1.19 shows a steel portal frame connected either by a pin base to its concrete foundation or alternatively where the design relies on moment fixity. In the former case (a) the foundation must be designed for the vertical and horizontal reactions whereas for the latter (b) its foundation must additionally resist bending moment. In general for portal frames the steelwork will be slightly heavier with pin bases but the foundations will be cheaper and less susceptible to movements of the subsoil.

For some structures it is vital to ensure that holding down bolts are capable of providing proper anchorage arrangement to prevent uplift under critical load conditions. An example is a water tower where uplift can occur at foundation level when the tank is empty under wind loading although the main design conditions for the tower members are when the tank is full.

1.7 Welding

1.7.1 Weld types

There are two main types of weld: *butt weld* and *fillet weld*. A butt weld (or groove weld) is defined as one in which the

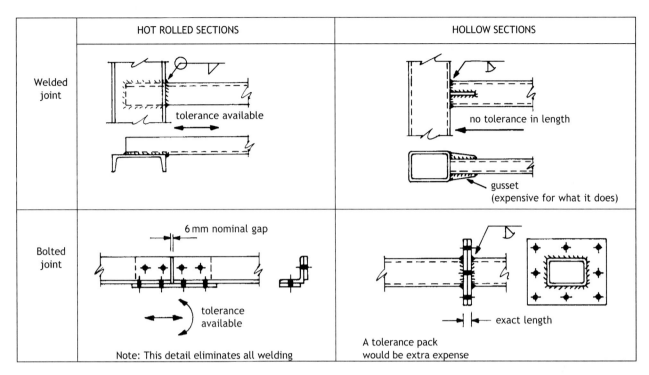

Figure 1.18 Connections in hot rolled and hollow sections.

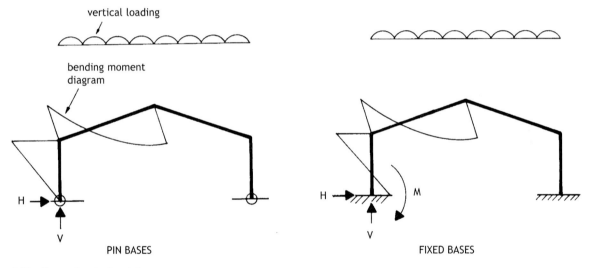

Figure 1.19 Connections to foundations.

metal lies substantially within the planes of the surfaces of the parts joined. It is able (if specified as a *full penetration butt weld*) to develop the strength of the parent material each side of the joint. A *partial penetration butt weld* achieves a specified depth of penetration only, where full strength of the incoming element does not need to be developed, and is regarded as a fillet weld in calculations of theoretical strength. Butt welds are shown in figure 1.20.

A fillet weld is approximately triangular in section formed within a re-entrant corner of a joint and not being a butt weld. Its strength is achieved through shear capacity of the weld metal across the throat, the weld size (usually) being specified as the leg length. Fillet welds are shown in figure 1.21.

1.7.2 Processes

Most workshops use electric arc manual (MMA), semiautomatic and fully automatic equipment as suited to the weld type and length of run. Either manual or semi-automatic processes are usual for short weld runs, with fully automatic welding being used for longer runs where the higher rates of deposition are less, being offset by extra set-up time. Detailing must allow for this. For example in fabricating a plate girder, full length web/flange runs are made first by

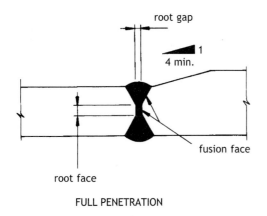
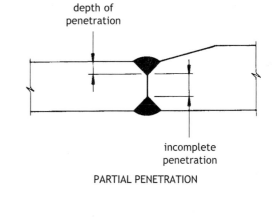

Figure 1.20   Butt welds showing double V preparations.

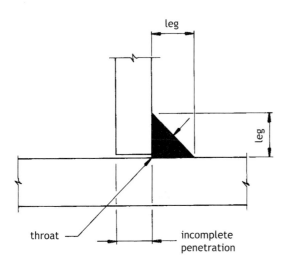
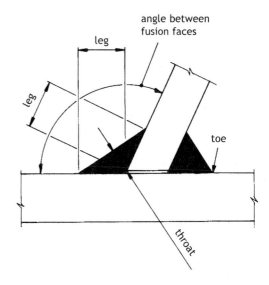

Figure 1.21   Fillet welds.

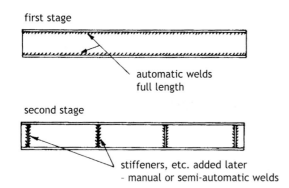

Figure 1.22   Sequence of fabrication.

automatic welding before stiffeners are placed with snipes to avoid the previous welding, as shown in figure 1.22.

Welding processes commonly used are shown in Table 1.9.

1.7.3 Weld size

In order to reduce distortion the *minimum* weld size consistent with *required* strength should be specified. The authors'

experience is that engineers tend to over-design welds in the belief that they are improving the product and they often specify butt welds when a fillet weld is sufficient. The result is a more expensive product which will be prone to unwanted distortion during manufacture. This can actually be detrimental if undesirable rectification measures are performed especially at site, or result in maintenance problems due to lack of fit at connections. An analogy exists in the art of the dressmaker who sensibly uses fine sewing thread to join seams to the thin fabric. The dressmaker would never use strong twine, far stronger, but which would tear out the edges of the fabric, apart from being unsightly and totally unnecessary.

Multiple weld-runs are significantly more costly than single run fillet welds and therefore joint design should aim for a 5 mm or 6 mm leg except for long runs, which will clearly be automatically welded when an 8 mm or 10 mm size may be optimum depending upon design requirements. For light fabrication using hollow sections with thickness 4 mm or less,

Table 1.9 Common weld processes.

Process	Automatic or manual	Shielding	Main use	Workshop or site	Comments	Maximum size fillet weld in single run
Manual metal arc (MMA)	Manual	Flux coating on electrode	Short runs }	Workshop or site	Fillet welds larger than 6 mm are usually multi-run, and are uneconomic	6 mm
Submerged arc (SUBARC)	Automatic	Powder flux deposited over arc and recycled	Long runs or heavy butt welds	Workshop or site	With twin heads simultaneous welds either side of joint are possible	10 mm
Metal inert gas (MIG)	Automatic or semi-automatic	Gas (generally carbon dioxide – CO_2)	Short or long runs	Workshop	Has replaced manual welding in many workshops. Slag is minimal so galvanised items can be treated directly	8 mm

then 4 mm size should be used where possible to reduce distortion and avoid burn-through. For thin platework (8 mm or less) the maximum weld size should be 4 mm and use of intermittent welds (if permitted) helps to reduce distortion. If it is to be hot-dip galvanised then distortion due to release of residual weld stresses can be serious if large welds are used with thin material. Intermittent welds should not be specified in exposed situations (because of corrosion risk) or for joints which are subject to fatigue loading such as crane girders, but are appropriate for internal areas of box girders and pontoons.

1.7.4 Choice of weld type

Butt welds, especially full penetration butt welds, should only be used where essential, such as in making up lengths of beam or girder flange into full strength members. Their high cost is due to the number of operations necessary, including edge preparation, back gouging, turning over, grinding flush (where specified) and testing, whereas visual inspection is often sufficient for fillet welds. Welding of end plates, gussets, stiffeners, bracings and web/flange joints should use fillet welds even if more material is implied. For example lapped joints should always be used in preference to direct butting, as shown in figure 1.23.

In the UK welding of structural steel is carried out to BS EN 1011 which requires weld procedures to be drawn up by the fabricator. It includes recommendations for any preheating of joints so as to avoid hydrogen induced cracking, this being sometimes necessary for high tensile steels. Fillet welds should where possible be returned around corners for

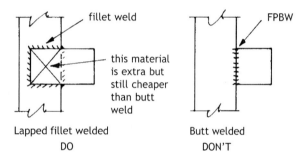

Figure 1.23 Welding using lapped joints.

a length of at least twice the weld size to reduce the possibility of failure emanating from weld terminations, which tend to be prone to start: stop defects.

1.7.5 Lamellar tearing

In design and detailing it should be appreciated that structural steels, being produced by rolling, possess different and sometimes inferior mechanical properties transverse to the rolled direction. This occurs because non-metallic manganese sulphides and manganese silica inclusions, which occur in steel making become extended into thin planar type elements after rolling. In this respect the structure of rolled steel resembles timber to some extent in possessing grain direction. In general this is not of great significance from a strength viewpoint. However, when large welds are made such that a fusion boundary runs parallel to the planar inclusion, the phenomenon of *lamellar tearing* can result. Such tearing is initiated and propagated by the considerable contractile stress across the thickness of the plate generated

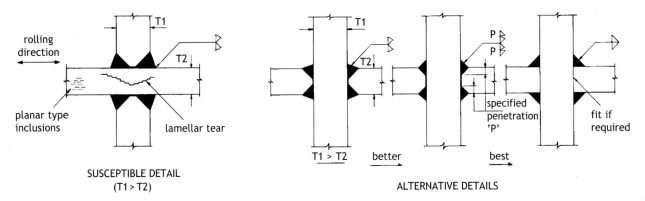

Figure 1.24 Lamellar tearing.

by the weld on cooling. If the joint is under *restraint* when welded, such as when a cruciform detail is welded which is already assembled as part of a larger fabrication then the possibility of lamellar tearing cannot be ignored. This is exacerbated where full penetration butt welds are specified not only because of the greater volume of weld metal involved, but because further transverse strains will be caused by the heat input of back-gouging processes used between weld runs to ensure fusion. The best solution is to avoid cruciform welds having full penetration butt welds. If cruciform joints are unavoidable then the thicker of the two plates should pass through, so that the strains which occur during welding are less severe. In other cases a special through thickness steel grade can be specified which has been checked for the presence of lamination type defects. However, the ultrasonic testing which is used may not always give a reliable guide to the susceptibility to lamellar tearing. Fortunately, most known examples have occurred during welding and have been repaired without loss of safety to the structure in service. However, repairs can be extremely costly and cause unforeseen delays. Therefore details which avoid the possibility of lamellar tearing should be used whenever possible. Figure 1.24 shows lamellar tearing together with suggested alternative details.

1.8 Bolting

1.8.1 General

Bolting is the usual method for forming site connections and is sometimes used in the workshop. The term 'bolt' used in its generic sense means the assembly of bolt, nut and appropriate washer. Bolts in clearance holes should be used except where absolute precision is necessary. *Black bolts* (the term for an untensioned bolt in a clearance hole 2 or 3 mm larger than the bolt dependent upon diameter) can

generally be used except in the following situations where slip is not permissible at working loads:

(1) Rigid connections – for bolts in shear.
(2) Impact-, vibration- and fatigue-prone structures, – e.g. silos, towers, bridges.
(3) Connections subject to stress reversal (except where due to wind loading only).

High-strength friction grip (HSFG) bolts should be used in these cases or, exceptionally, precision bolts in close tolerance holes ($+0.15\,mm-0\,mm$) may be appropriate.

If bolts of different grade or type are to be used on the same project then it is wise to use different diameters. This will overcome any possible errors at the erection stage and prevent incorrect grades of bolt being used in the holes. For example, a typical arrangement would be:

All grade 4.6 bolts – 20 mm diameter
All grade 8.8 bolts – 24 mm diameter

Previous familiar bolting standards BS 3692 and BS 4190 have been replaced by a range of European standards (EN 24014, 24016–24018, 24032 and 24034). Whilst neither the old nor the new standards include the term 'fully threaded bolts', they do permit their use. Bolt manufacturers have been supplying fully threaded bolts for some time to the increasing number of steelwork contractors using them as the normal structural fastener in buildings. They are ordinary bolts in every respect except that the shank is threaded for virtually its full length. This means that a more rationalised and limited range of bolt lengths can be used. The usual variable of bolt length (grip + nut depth + washer + minimum thread projection past the nut) can be replaced by a variable projection beyond the tightened nut. This has a significant effect on the number of different bolt lengths required.

BLACK BOLTS   pretension   HSFG BOLTS

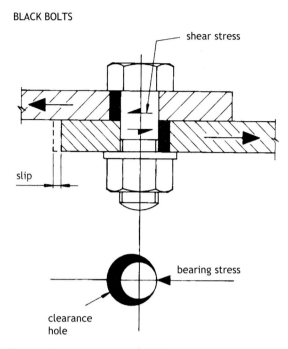

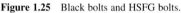

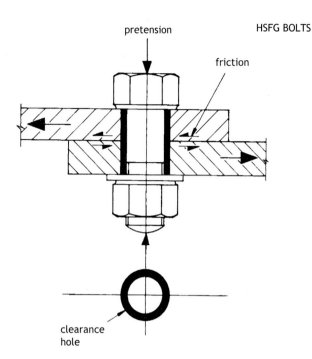

Figure 1.25 Black bolts and HSFG bolts.

Although the new European standards have been published, their adoption by the industry has been a slow process. Bolt manufacturers still continue to produce bolts, nuts and washers in compliance with the existing British standards. It is for this reason that the technical information relating to bolting in this manual refers generally to the relevant British standard.

Black bolts and HSFG bolts are illustrated in figure 1.25. The main bolt types available for use in the UK are shown in Table 1.10.

The European continent system of strength grading introduced with the ISO system is given by two figures, the first being one tenth of the minimum ultimate stress in kgf/mm^2 and the second is one tenth of the percentage of the ratio of minimum yield stress to minimum ultimate stress. Thus '4.6 grade' means that the minimum ultimate stress is 40 kgf/mm^2 and the yield stress 60 per cent of this. The yield stress is obtained by multiplying the two figures together to give 24 kgf/mm^2. For higher tensile products where the yield point is not clearly defined, the stress at a permanent set limit is quoted instead of yield stress.

The single grade number given for nuts indicates one tenth of the proof load stress in kgf/mm^2 and corresponds with the bolt ultimate strength to which it is matched, e.g. an 8 grade nut is used with an 8.8 grade bolt. It is permissible to use a higher strength grade nut than the matching bolt number and grade 10.9 bolts are supplied with grade 12 nuts since grade 10 does not appear in the British Standard series.

Table 1.10 Bolts used in UK.

Type	BS No	Main use	Workshop or site
Black bolts, grade 4.6 (mild steel)	BS 4190 (nuts and bolts) BS 4320 (washers)	As black bolts in clearance holes	Workshop or site
High tensile bolts, grade 8.8	BS 3692 (nuts and bolts) BS 4320 (washers)	As black bolts in clearance holes As precision bolts in close tolerance holes	Workshop or site Workshop
HSFG bolts, general grade	BS 4395 Pt 1 (bolts, nuts and washers)	Bolts in clearance holes where slip not permitted. Used to BS 4604 Pt 1	Workshop or site
Higher grade	BS 4395 Pt 2 (bolts, nuts and washers)	Bolts in clearance holes where slip not permitted. Used to BS 4604 Pt 2	Workshop or site
Waisted shank	BS 4395 Pt 3 (bolts, nuts and washers)	Bolts in clearance holes where slip not permitted. Used to BS 4604 Pt 3	Little used

To minimise risk of thread stripping at high loads, BS 4395 high strength friction grip bolts are matched with nuts one class higher than the bolt.

1.8.2   High strength friction grip (HSFG) bolts

A pre-stress of approximately 70 per cent of ultimate load is induced in the shank of the bolts to bring the adjoining plies into intimate contact. This enables shear loads to be transferred by friction between the interfaces and makes for rigid connections resistant to movement and fatigue. HSFG bolts thus possess the attributes possessed by rivets, which welding and bolts displaced during the early 1950s.

During tightening the bolt is subjected to two force components:

(1)   The induced axial tension.
(2)   Part of the torsional force from the wrench applied to the bolt via the nut thread.

The stress compounded from these two forces is at its maximum when tightening is being completed. Removal of the wrench will reduce the torque component stress, and the elastic recovery of the parts causes an immediate reduction in axial tension of some 5 per cent followed by further relaxation of about 5 per cent, most of which takes place within a few hours. For practical purposes, this loss is of no consequence since it is taken into account in the determination of the slip factor, but it illustrates that a bolt is tested to a stress above that which it will experience in service. It may be said that if a friction grip bolt does not break in tightening the likelihood of subsequent failure is remote. The bolt remains in a state of virtually constant tension throughout its working life. This is most useful for structures subject to vibration, e.g. bridges and towers. It also ensures that nuts do not become loose with risk of bolt loss during the life of the structure, thus reducing the need for continual inspection.

Mechanical properties for general grade HSFG bolts (to BS 4395: Part 1) are similar to grade 8.8 bolts for sizes up to and including M24. Although not normally recommended, grade 8.8 bolts can exceptionally be used as HSFG bolts.

HSFG bolts may be tightened by three methods, viz:

(1)   Torque control
(2)   Part turn method
(3)   Direct tension indication.

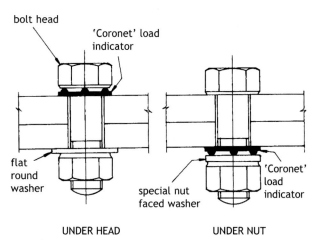

Figure 1.26   Use of 'Coronet' load indicator.

The latter is now usual practice in the UK and the well-established 'Coronet'* load indicator has often been used which is a special washer with arched protrusions raised on one face. It is normally fitted under the standard bolt head with the protrusions facing the head, thus forming a gap between the head and load indicator face. On tightening, the gap reduces as the protrusions depress and when the specified gap (usually 0.40 mm) is obtained, the bolt tension will not be less than the required minimum. Assembly is shown in figure 1.26.

1.8.3   Tension control bolts

Tension control bolts, or TCBs as they are commonly known, are replacing conventional HSFG bolts simply because they are very quick and easy to install using lightweight electrical shear wrenches. Guaranteed tension together with visual inspection provides engineers with the assurance that connections are tightened in accordance with specifications.

High strength TCBs are used in a wide range of applications from bridge splice plates to beam to column connections, from stadia roof trusses to rail switches and crossings. The combination of superior tensile strength together with phenomenal ductility results in a universal bolt that can be employed in most steelwork connections.

TCBs have a domed head and the threaded section of the bolt is extended to form a waisted portion and a splined end. The TCB assembly is completed with a tough hardened flat washer fitted under the nut. Tightening is achieved with

*'Coronet' load indicators are manufactured by Cooper & Turner Limited, Vulcan Works, Vulcan Road, Sheffield S9 2FW, UK.

the aid of a shear wrench with a socket that locates on the nut and spline. When the correct shank tension is reached the spline is sheared, giving instant and visible inspection. Although the bolt is properly tightened and is resistant to any subsequent vibration, it can be loosened and removed by conventional methods.

TCBs can be installed from either side of the work to accommodate any access limitations. Only a flat washer is used, under the nut, and no other load indicating device is required. TCBs are tightened from one face of the work without the need to hold the bolt head.

1.8.4   European bolting standards

The launch of the new European standards for design (the Eurocodes) and fabrication (BS EN 1090-2) of structural steelwork is associated with the introduction of a set of European standards for non-preloadable (ordinary) and preloadable (high strength friction grip) bolts.

This is a brief description of the different types of European pre-loadable bolts and the major issues that are likely to be encountered when using these bolts.

In Europe there are two approaches to achieving the necessary ductility in preloaded bolt, nut, and washer assemblies, therefore in developing the series of European product standards, BS EN 14399, it was agreed to develop two parallel systems. The HR (British/French) and the HV (German) systems reflect these two approaches and the differences between the two are explained below. With both types of bolt, the fact that the thread may be subject to plastic strains during tightening means that bolts and nuts that have been fully preloaded must not be re-used if removed.

HR (British/French) bolt

The British/French approach following BS EN 14399-3 and BS 4395 is to use thick nuts and long thread lengths in the bolt assembly to obtain ductility predominantly by plastic elongation of the bolt. The longer thread length is necessary to ensure that the induced strain is not localised. These bolts are relatively insensitive to over-tightening during preloading, although suite control is still important. Furthermore, if severely over-tightened during preloading the ductile failure mode of the bolt assembly is predominantly by bolt breakage, which is readily detectable.

HV (German) bolt

The German approach following BS EN 14399-4 and DIN 6914 is to use thinner nuts and shorter thread lengths to obtain the required ductility by plastic deformation of the threads within the nut. In Germany, the HV bolt assembly is used in both preloaded and non-preloaded applications, and it can be argued that in the event of failure by thread plastic deformation the assembly still acts as a non-preloaded assembly. These assemblies are more sensitive to over-tightening during preloading and therefore require more site control. If severely over-tightened during preloading the mode of failure by plastic deformation of the engaged thread of the bolt assembly offers little indication of impending failure.

Marking

It is vital to avoid mixing up the components of both systems and this is not helped by the same standard covering both types of bolt. Bolts and nuts for both systems are standardised in separate parts of the product standard BS EN 14399 and clearly marked as components for the separate systems. Bolts and nuts from the same system will be stamped with their system designation, HR or HV, in order to avoid confusion. In addition, bolts and nuts will be stamped with their property class (i.e. grade 8.8 or 10.9 for bolts and 8 or 10 for nuts as appropriate). For the HR system the following possibilities exist:

- Bolts to class HR 8.8 with nuts to class HR 8, or HR 10
- Bolts to class HR 10.9 with nuts to class HR 10.

The HR 8.8 bolt is very similar (in dimensions and properties) to the Part 1 general grade HSFG bolt to BS 4395 and likewise the HR 10.9 bolt is very similar to the Part 2 higher grade HSFG bolt to BS 4395.

Key Points

(1)   There are two types of preloadable bolt assembly, the British/French HR bolts covered by BS EN 14399-3 and the German HV assembly covered by BS EN 14399-4.

(2)   The HR assembly is similar to the BS 4395 bolt and is generally less sensitive to over-tightening.

(3)   The HV assembly is more sensitive to over-tightening and requires more control on site.

(4)   Components from both types of bolt assembly must not be mixed up.

(5) Three methods of tightening are given in the European fabrication standard BS EN 1090-2: torque control, 'combined' and direct tension indicator.

(6) The requirements for CE marking of preloadable fasteners are given in BS EN 14399-1.

1.9 Dos and don'ts

The overall costs of structural steelwork are made up of a number of elements which may vary considerably in proportion depending upon the type of structure and site location. However a typical split is shown in Table 1.11.

Table 1.11 Typical cost proportion of steel structures.

	Materials %	Workmanship %	Total %
Materials	30	0	30
Fabrication	0	45	45
Erection	0	15	15
Protective treatment	5	5	10
Total	35	65	100

It may be seen that the materials element (comprising rolled steel from the mills, bolts, welding consumables, paint and so on) is significant, but constitutes considerably less in proportion than the workmanship. This is why the economy of steel structures depends to a great extent on *details* which allow easy (and therefore less costly) fabrication and erection. Minimum material content is important in that designs should be efficient, but more relevant is the correct selection of structural type and fabrication details. The use of automated fabrication methods has enabled economies to be made in overall costs of steelwork, but this can only be realised fully if details are used which permit tolerance (see section 1.4) so that time consuming (and therefore costly) rectification procedures are avoided at site. Often if site completion is delayed then severe penalties are imposed on the steel contractor and this affects the economy of steelwork in the long term.

For this reason one of the purposes of this manual is to promote the use of details which will avoid problems both during fabrication and erection. Figures 1.27 and 1.28 show a series of dos and don'ts which are intended to be used as a general guide in avoiding uneconomic details. Figure 1.29 gives dos and don't related to corrosion largely so as to permit maintenance and avoid moisture traps.

1.10 Protective treatment

When exposed to the atmosphere all construction materials deteriorate with time. Steel is affected by atmospheric corrosion and normally requires a degree of protection, which is no problem but requires careful assessment depending upon:

- Aggressiveness of environment
- Required life of structure
- Maintenance schedule
- Method of fabrication and erection
- Aesthetics.

It should be remembered that for corrosion to occur air and moisture both need to be present. Thus, permanently embedded steel piles do not corrode, even though in contact with water, provided air is excluded by virtue of impermeability of the soil. Similarly, the internal surfaces of hollow sections do not corrode provided complete sealing is achieved to prevent continuing entry of moist air.

There is a wide selection of protective systems available, and that used should adequately protect the steel at the most economic cost. Detailing has an important influence on the life of protective treatment. In particular, details should avoid the entrapment of moisture and dirt between profiles or elements, especially for external structures. Figure 1.29 gives dos and don'ts related to corrosion. Provided that the ends are sealed by welding, then hollow sections do not require treatment internally. For large internally stiffened hollow members which contain internal stiffening, such as *box girder bridges* and *pontoons* needing future inspection, it is usual to provide an internal protective treatment system. Access manholes should be sealed by covers with gaskets to prevent ingress of moisture as far as possible, allowing use of a cheaper system. For immersed structures such as pontoons, which are inaccessible for maintenance, corrosion prevention by cathodic protection may be appropriate.

Adequate preparation of the steel surface is of the utmost importance before application of any protective system. Modern fabricators are properly equipped in this respect such that the life of systems has considerably extended. For external environments it is especially essential that all millscale is removed which forms when the hot surface of rolled steel reacts with air to form an oxide. If not removed it will eventually become detached through corrosion. Blast

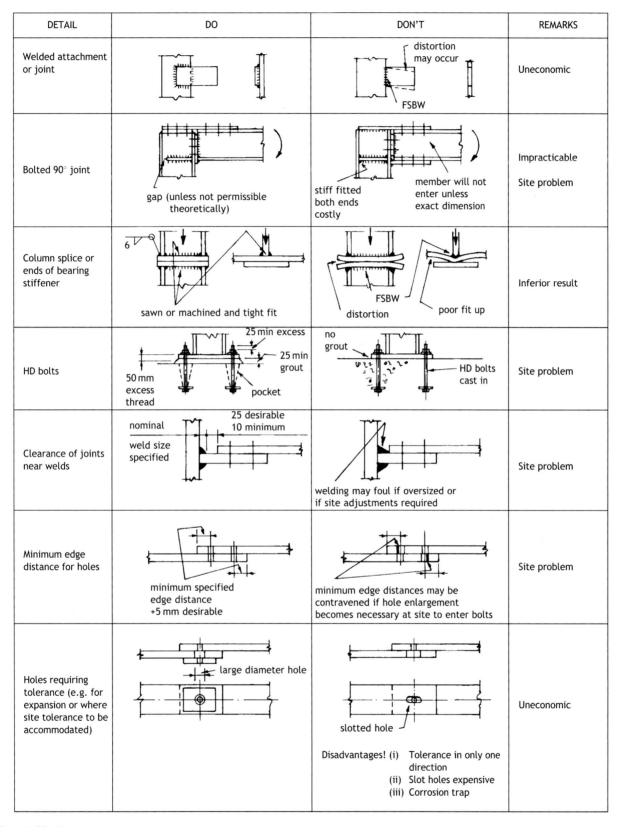

Figure 1.27 Dos and don'ts.

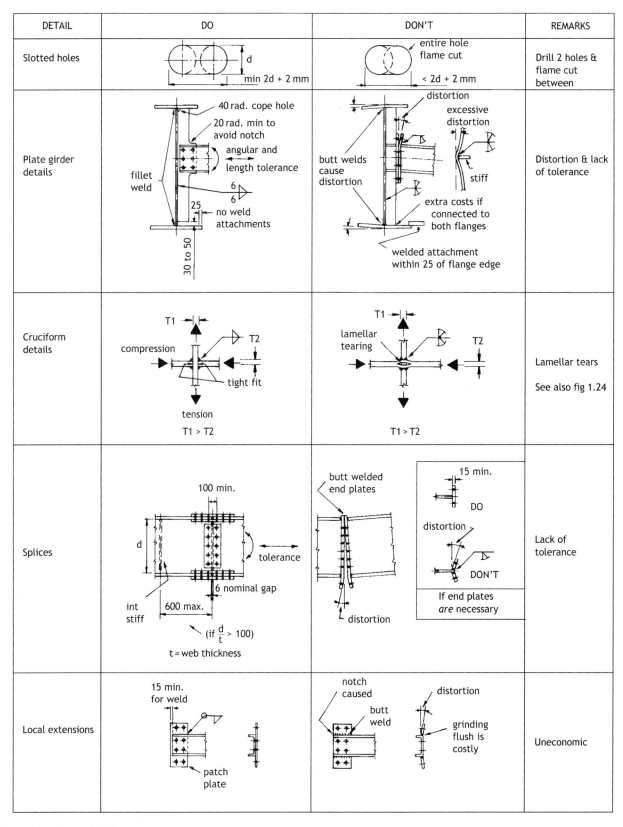

Figure 1.28 Dos and don'ts.

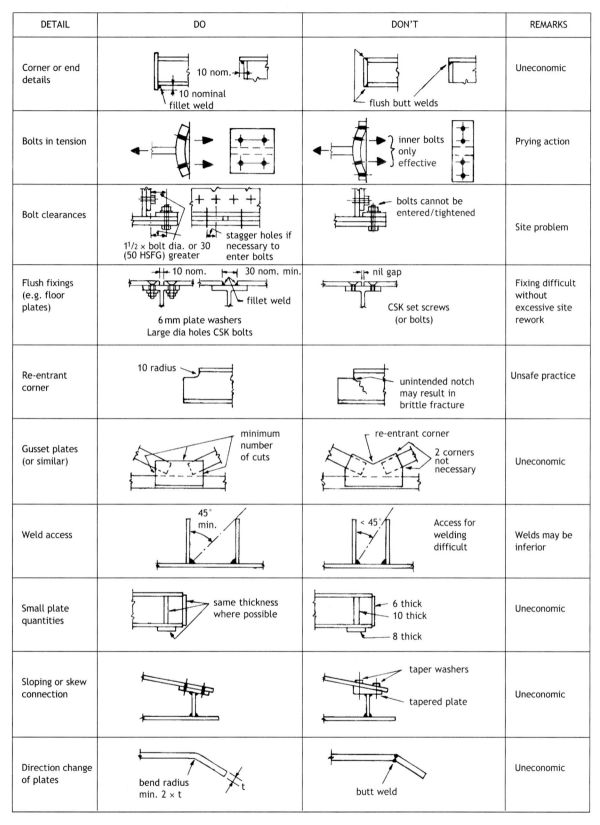

DETAIL	DO	DON'T	REMARKS
Corner or end details	10 nom. 10 nominal fillet weld	flush butt welds	Uneconomic
Bolts in tension		inner bolts only effective	Prying action
Bolt clearances	$1\frac{1}{2}$ × bolt dia. or 30 (50 HSFG) greater / stagger holes if necessary to enter bolts	bolts cannot be entered/tightened	Site problem
Flush fixings (e.g. floor plates)	10 nom. 30 nom. min. fillet weld 6 mm plate washers Large dia holes CSK bolts	nil gap CSK set screws (or bolts)	Fixing difficult without excessive site rework
Re-entrant corner	10 radius	unintended notch may result in brittle fracture	Unsafe practice
Gusset plates (or similar)	minimum number of cuts	re-entrant corner 2 corners not necessary	Uneconomic
Weld access	45° min.	< 45° Access for welding difficult	Welds may be inferior
Small plate quantities	same thickness where possible	6 thick 10 thick 8 thick	Uneconomic
Sloping or skew connection		taper washers tapered plate	Uneconomic
Direction change of plates	bend radius min. 2 × t	butt weld	Uneconomic

Figure 1.28 *Contd*

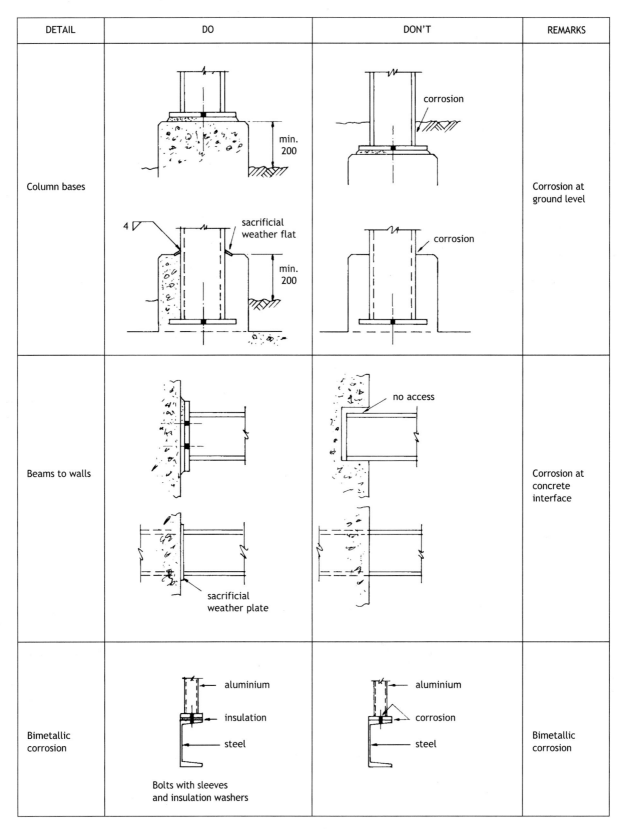

DETAIL	DO	DON'T	REMARKS
Column bases			Corrosion at ground level
Beams to walls			Corrosion at concrete interface
Bimetallic corrosion			Bimetallic corrosion

Figure 1.29   Dos and don'ts – corrosion.

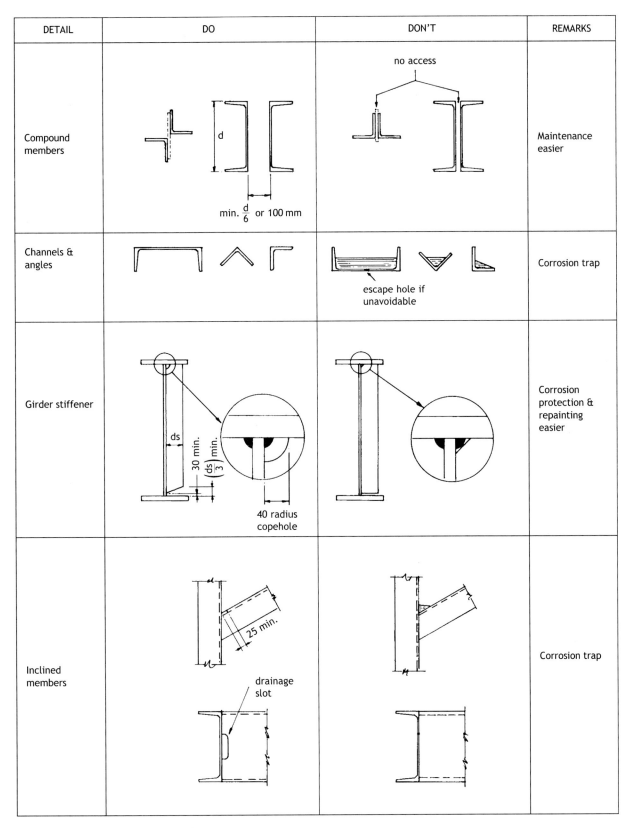

DETAIL	DO	DON'T	REMARKS
Compound members	min. $\frac{d}{6}$ or 100 mm	no access	Maintenance easier
Channels & angles		escape hole if unavoidable	Corrosion trap
Girder stiffener	ds 30 min. $(\frac{ds}{3})$ min. 40 radius copehole		Corrosion protection & repainting easier
Inclined members	25 min. drainage slot		Corrosion trap

Figure 1.29 *Contd*

cleaning is widely used to prepare surfaces, and other processes such as hand cleaning are less effective, although acceptable in mild environments. Various national standards for the quality of surface finish achieved by blast cleaning are correlated in Table 1.12.

Table 1.12 National standards for grit blasting.

British Standard	Swedish Standard	USA Steel Structures Painting Council
BS 7079	SIS 05 59 00[2]	SSPC[3]
1st quality	Sa 3	White metal
2nd quality	Sa $2^1/_2$	Near white
3rd quality	Sa 2	Commercial

A brief description is given for a number of accepted systems in Table 1.13 based on UK conditions to BS EN ISO 14713 and Department of Transport guidance.[4] Specialist advice may need to be sought in particular environments or areas.

The following points should be noted when specifying systems:

(1) Metal coatings such as hot dip galvanizing and aluminium spray give a durable coating more resistant to site handling and abrasion but are generally more costly.

(2) Hot dip galvanizing is not suitable for plate thicknesses less than 5 mm. Welded members, especially if slender, are liable to distortion due to release of residual stress and may need to be straightened. Hot dip galvanizing is especially suitable for piece-small fabrications which may be vulnerable to handling damage, such as when despatched overseas. Examples are towers or lattice girders with bolted site connections.

(3) Most sizes and shapes of steel fabrications can be hot dip galvanized, but the dimensions of the galvanizing bath determine the size and shape of articles that can be coated in a particular works. Indicative UK maximum single dip sizes (length, depth, width) of assemblies are:

> $20.0 \times 1.45 \times 2.7$ m
> $7.5 \times 2.0 \times 3.25$ m
> $5.75 \times 1.9 \times 3.5$ m

However, articles which are larger than the bath dimensions can by arrangement sometimes be galvanized by double-dipping. Although generally it is preferable to process work in a single dip, the corrosion protection afforded through double-dipping is no different from that provided in a single dip. Sizes of articles which can be double-dipped should always be agreed with the galvanizers. By using double-dipping UK galvanizing companies can now handle lengths up to 29.0 m or widths up to 4.8 m.

(4) For HSFG bolted joints the interfaces must be grit blasted to Sa$2^1/_2$ quality or metal sprayed only, without any paint treatment to achieve friction. A reduced slip factor must be assumed for galvanized steelwork. During painting in the workshop the interfaces are usually masked with tape, which is removed at site assembly. Paint coats are normally stepped back at 30 mm intervals, with the first coat taken 10 to 15 mm inside the joint perimeter. Sketches may need to be prepared to define painted/masked areas.

(5) For non friction bolted joints the first two workshop coats should be applied to the interfaces.

(6) Micaceous iron oxide paints are obtainable in limited colour range only (e.g. light grey, dark grey, silver grey) and provide a satin finish. Where a decorative or gloss finish is required then another system of overcoating must be used.

(7) Surfaces in contact with concrete should be free of loose scale and rust but may otherwise be untreated. Treatment on adjacent areas should be returned for at least 25 mm and any metal spray coating must be overcoated.

(8) Treatment of bolts at site implies blast cleaning unless they have been hot dip galvanized. As an alternative, consideration can be given to use of electro-plated bolts, degreased after tightening followed by etch priming and painting as for the adjacent surfaces.

(9) Any delay between surface preparation and application of the first treatment coat must be kept to the absolute minimum.

(10) Lifting cleats should be provided for large fabrications exceeding say 10 tonnes in weight to avoid handling damage.

(11) The maximum amount of protective treatment should be applied at workshop in enclosed conditions. In some situations it would be advisable to apply at least the final paint coat at site after making good any erection damage.

1.11 Drawings

1.11.1 Engineer's drawings

Engineer's drawings are defined as the drawings which describe the employer's requirements and main details.

Table 1.13a Typical protective treatment systems for building structures.

Steelwork location	Environment category	Time	Structure type	Surface preparation	Metal coating	Paint coats 1	2	3	4	Total paint dry film thickness	Treatment of bolts
Interior	A	20–35 years	Buildings	Sa $2\frac{1}{2}$	–	HB zpa	HB alkyd finish	–	–	120	Paint
Interior	B	15 years	Buildings	Sa $2\frac{1}{2}$	–	zre	CR alkyd	HB CR finish	–	135	Paint
Interior	C	20 years	Buildings	Sa 3	–	ezs	CR alkyd	HB CR finish	–	185	Paint
Exterior	D	10–15 years	General	Sa $2\frac{1}{2}$	–	HB zpa	HB alkyd finish	–	–	135	Paint
Exterior	E	12–18 years	General	Sa $2\frac{1}{2}$	–	zre	HB zp	Modified alkyd MIO	–	200	Paint
Exterior	F	16–24 years	Piece-small fabrication	–	Hot dip galvanize	–	–	–	–	85 (min)	Galvanize
Exterior	G	16–24 years	General	Sa 3	–	ezs	CR alkyd	HB MIO CR	HB MIO CR	260	Galvanize

Notes to Table 1.13a

1. Environments:

    A – dry heated interiors
    B – interiors subject to occasional condensation
    C – interiors subject to frequent condensation
    D – normal inland
    E – polluted inland
    F – normal coastal
    G – polluted coastal

2. Time indicated is approximate period in years to first major maintenance. The time will be subject to variation depending upon the micro-climate around the structure. Maintenance may need to be more frequent for decorative appearance.

3. The number of coats given is indicative. A different number of coats may be necessary depending upon the method of application in order to comply with the dry film thickness specified.

Abbreviations for paint coats:

CR – chlorinated rubber
ezs – ethyl zinc silicate
HB – high build
MIO – micaceous iron oxide
zp   – zinc phosphate
zpa – zinc phosphate modified alkyd
zre – zinc rich epoxy (2 pack)
‖   – despatch to site and erect

Table 1.13b Summary table of Highways Agency painting specifications for Highway Works Series 1900 – 8th edition (1998).

Environment location	Access	System type	Metal spray	1st coat	2nd coat	3rd coat	4th coat	5th coat	6th coat	Minimum total dry film thickness of paint system (μm)
Inland	R	4	–	Zinc phosphate AR blast primer	Zinc phosphate AR undercoat	Zinc phosphate AR undercoat	MIO AR undercoat	AR finish	–	250
Inland	R	4 Alt	–	Zinc phosphate HB/epoxy primer QD	MIO, HB QD epoxy undercoat	Polyurethane (2 pack) finish or MC polyurethane finish	–	–	–	300
Marine	R	8	–	Zinc phosphate AR blast primer	Zinc phosphate AR undercoat	Zinc phosphate AR undercoat	MIO AR undercoat	AR undercoat	AR finish	300
Marine	R	8 Alt	–	Zinc phosphate HB epoxy primer	MIO, HB QD epoxy undercoat	Polyurethane (2 pack) finish or MC polyurethane finish	–	–	–	300
Marine	D	10	Aluminium metal spray (100 μm)	Aluminium epoxy sealer	Zinc phosphate AR undercoat	MIO AR undercoat	MIO AR undercoat	AR finish	–	250
Marine	D	10 Alt	Aluminium metal spray (100 μm)	Aluminium epoxy sealer	Zinc phosphate HB QD epoxy undercoat	MIO HB QD epoxy undercoat	Polyurethane (2 pack) finish or MC polyurethane finish	–	–	300
Interiors of box girders	R or D	11	–	Zinc phosphate HB QD epoxy primer	MIO HB QD epoxy finish	–	–	–	–	200
Bridge parapets	R or D	13	Galvanize	T-wash	Zinc phosphate AR undercoat	MIO AR undercoat	MIO AR undercoat	–	–	150

Notes to Table 1.13b

1. Environments: Location of structures. Two locations are considered: 'Inland' and 'Marine'.
Structures out of reach of sea salt spray are considered as being 'Inland'.
Structures which can be affected by sea salt spray are considered as being 'Marine'.

2. Required durability: For the basic systems (except for lighting columns), the periods which will be sufficiently accurate for both access situations and the environments described above are:
no maintenance up to 12 years
minor maintenance from 12 years
major maintenance after 20 years.

3. Standards of surface preparation quality and finish should relate to cleanliness (e.g. BS 7079 Part A1/ISO 8501-1) and profile (e.g. BS 7079 Part C to C4/ISO 8503-1 to 4).

4. Details given are based on UK Highways Agency 'Specification for Highway Works', Volume 1 – Series 1900 Protection of Steelwork Against Corrosion, and the accompanying 'Notes for Guidance on the Specification for Highway Works', Volume 2 – Series NG 1900.

Abbreviations for paint coats:

D   – Difficult
R   – Ready
AR – Acrylated Rubber
MIO – Micaceous Iron Oxide
HB – High Build
QD – Quick Drying
MC – Moisture-Cured
‖   – Despatch to site and erect

Usually they give all leading dimensions of the structure including alignments, levels, clearances, member size and show steelwork *in an assembled form*. Sometimes, especially for buildings, connections are not indicated and must be designed by the fabricator to forces shown on the engineer's drawings requiring submission of calculations to the engineer for approval. For major structures such as bridges the engineer's drawings usually give details of connections including sizes of all bolts and welds. Most example drawings of typical structures included in this manual can be defined as engineer's drawings.

Engineer's drawings achieve the following purposes:

(1) Basis of engineer's cost estimate before tenders are invited.
(2) To invite tenders upon which competing contractors base their prices.
(3) Instructions to the contractor during the contract (i.e. *contract drawings*), including any revisions and variations. Most contracts usually involve revisions at some stage due to the employer's amended requirements or due to unexpected circumstances such as variable ground conditions.
(4) Basis of measurement of completed work for making progressive payments to the contractor.

1.11.2 Workshop drawings

Workshop drawings (or shop details) are defined as the drawings prepared by the steelwork contractor (i.e. the fabricator, often in capacity of a subcontractor) showing each and every component or member in full detail for fabrication. A requirement of most contracts is that workshop drawings are submitted to the engineer for approval, but that the contractor remains responsible for any errors or omissions. Most responsible engineers nevertheless carry out a detailed check of the workshop drawings and point out any apparent shortcomings. In this way any undesirable details are hopefully discovered before fabrication and the chance of error is reduced. Usually a marked copy is returned to the contractor who then amends the drawings as appropriate for re-submission. Once approved the workshop drawings should be correctly regarded as contract drawings.

Workshop drawings are necessary so that the steelwork contractor can organise efficient production of large numbers of similar members, but with each having slightly different details and dimensions. Usually each member is shown fabricated as it will be delivered on site. Confusion

and errors can be caused under production conditions if only typical drawings showing many variations, lengths and 'opposite handing' for different members are issued. *Workshop drawings of members must include reference dimensions to main grid lines* to facilitate cross referencing and checking. This is difficult to undertake without the possibility of errors if members are drawn only in isolation. All extra welds or joints necessary to make up member lengths must be included on workshop drawings. Marking plans must form part of a set of workshop drawings to ensure correct assembly and to assist planning for production, site delivery and erection. A General Arrangement drawing is often also required, giving overall setting out including holding down bolt locations from which workshop drawing lengths, skews and connections have been derived. Often the engineer's drawings are inadequate for this purpose because only salient details and overall geometry will have been defined.

Workshop drawings must detail camber geometry for girders so as to counteract (where required and justified) dead load deflection, including the correct inclinations of bearing stiffeners. For site welded connections the workshop drawings must include all temporary welding restraints for attachment and joint root gap dimensions allowing for predicted weld shrinkage. Each member must be allocated a mark number. A system of 'material marks' is also usual and added to the workshop drawings so that each stiffener or plate can be identified and cut by the workshop from a material list.

1.11.3 Computer aided detailing

Reference should be made to Chapter 6 *Computer Aided Detailing* for a review of the wide use of CAD by engineers and steelwork contractors to improve their efficiency and minimise costly errors in their workshop fabrication processes and site construction activities.

1.12 Codes of practice

In the UK appropriate UK and other European Standards for the design and construction of steelwork are as summarised below. The introduction of the new European standards has led in recent years to a great deal of discussion and varied interpretation of the design methods which should be used for new structures to be built in the UK or which are designed by British firms for construction overseas.

Currently some of these new standards – or Eurocodes – are used alongside the existing UK Codes of Practice for design and construction. In the structural Eurocodes, certain safety related numerical values such as partial safety factors, are only indicative. The values to be used in practice have been left to be fixed by the national authorities in each country and published in the relevant National Annex (NA). These values, referred to as 'boxed' values, which are used for buildings to be constructed in the UK are set down in the UK NA, which is bound in with the European CEN text of the relevant Eurocode.

The NA also specifies the loading codes to be used for steel structures constructed in the UK, pending the availability of harmonised European loading information in the Eurocodes. It also includes additional recommendations to enable the relevant Eurocode to be used for the design of structures in the UK. The relevant NA should always be consulted for buildings to be constructed in any other country. Different design criteria may need to be applied for example in the cases of varied loadings, earthquake effects, temperature range and so on.

1.12.1   Buildings

Steelwork in buildings is designed and constructed in the UK to BS 5950. The revised Part 1 published in 2001 is a Code for the design of hot rolled sections in buildings. A guide is available[5] giving member design capacities, together with those for bolts and welds. BS 5950 Part 2 is a specification for materials fabrication and erection, and BS EN ISO 14713 gives guidance on protective treatment. BS 5950 Part 5 deals with cold formed sections.

BS 5950 uses the *limit state* concept in which various limiting states are considered under factored loads. The main limit states are:

Ultimate limit state	Serviceability limit state
Strength (i.e. collapse)	Deflection
Stability (i.e. overturning)	Vibration
Fatigue fracture	Repairable fatigue damage
Brittle fracture	Corrosion

The following must be satisfied:

$$\text{Specified loads} \times \gamma f (\text{load factor}) \le \frac{\text{Material strength}}{\gamma m \text{ (material factor)}}$$

where $\gamma m = 1.0$

Values of the load factor are summarised in Table 1.14.

Table 1.14   BS 5950 load factors γ f and combinations.

Loading	Load factor γ f
Dead load	1.4
Dead load restraining uplift or overturning	1.0
Dead load acting with wind and imposed loads combined	1.2
Imposed loads	1.6
Imposed load acting with wind load	1.2
Wind load	1.4
Wind load acting with imposed load or crane load	1.2
Forces due to temperature effects	1.2
Crane loading effects	
Vertical load	1.6
Vertical load acting with horizontal loads (crabbing or surge)	1.4
Horizontal load	1.6
Horizontal load acting with vertical load	1.4
Crane load acting with wind load*	1.2

*When considering wind or imposed load and crane loading acting together the value of γ f for dead load may be taken as 1.2.

For the ultimate limit state of fatigue and all serviceability limit states γ f = 1.0.

In this manual any load capacities give are in the terms of BS 5950 *ultimate* strength (i.e. material strength $\gamma m = 1.0$), generally a function of the guaranteed yield stress of the material from EN material standards. They must be compared with factored working loads as given by Table 1.14 in satisfying compliance. If a working load is supplied then its appropriate proportions should be multiplied by the load factors from Table 1.14. As an approximation a working load can be multiplied by an averaged load factor of say 1.5 if the contributions of dead and imposed loads are approximately equal.

BS EN 1993-1 'Eurocode 3: Design of Steel Structures: Part 1.1 General Rules for Buildings (EC3)' sets out the principles for the design of all types of steel structures as well as giving design rules for buildings. The transition from BS 5950 to EC3 is inevitably a slow process and, for the present, at least, both these two design standards will be used by UK designers.

1.12.2   Bridges

Bridges are designed and constructed to BS 5400 which covers steel, concrete, composite construction, fatigue, and bearings. It is adopted by the main UK highway and railway bridge authorities. It has been widely accepted in other countries and used as a model for other Codes. The UK Highways Agency implements BS 5400 with its own standards which in some cases vary with individual Code clauses. In particular the intensity of highway loading is increased to reflect the higher proportion of heavy commercial vehicles using UK highways since publication of the code.

BS 5400 uses a limit state concept similar to BS 5950. Many of the strength formulae are similar but there are additional clauses dealing with, for example, longitudinally stiffened girders, continuous composite beams and fatigue. In BS 5400 the breakdown of partial safety factors and the assessment of material strengths are different so that any capacities given in this book, where applicable to bridges, should not be used other than as a rough guide.

BS EN 1993-2: 1997 Eurocode 3: Design of Steel Structures: Part 2: Steel Bridges sets out the principles for the design of most types of steel road and railway bridges as well as giving design rules for the steel parts of composite bridges. For the design of steel and concrete composite bridges BS EN 1994-2: Eurocode 4: Part 2 will provide the future design rules. Like building structures, the transition from BS 5400 to EC3: Part 2 and EC4: Part 2 is inevitably a slow process and, for the present, at least, all of these design standards will be used in the appropriate circumstances by UK designers.

2 Detailing Practice

2.1 General

Drawings of steelwork, whether engineer's drawings or workshop drawings, should be carried out to a uniformity of standard to minimise the possible source of errors. Present day draughting practice is predominantly to use computer aided detailing systems, although in some situations traditional drawing board methods are still used. Whichever methods are used, individual companies will have particular requirements suited to their own operation, but the guidance given here is intended to reflect good practice. Certain conventions such as welding symbols are established by a standard or other codes and should be used wherever possible.

2.2 Layout of drawings

Drawing sheet sizes should be standardised. BS EN ISO 4157 gives the international 'A' series, but many offices use the 'B' series. Typical sizes used are shown in Table 2.1.

Table 2.1 Drawing sheet sizes.

Designation	Size mm	Main purpose
A0*	1189 × 841	Arrangement drawings
A1*	841 × 594	Detailed drawings
A2	594 × 420	Detailed drawings
A3*	420 × 297	Sketch sheets
A4*	297 × 210	Sketch sheets
B1	1000 × 707	Detailed drawings

*Widely used.

All drawings must contain a title block including company name, columns for the contract name/number, client, drawing number, drawing title, drawn/checked signatures, revision block, and notes column. Notes should, as far as possible, all be in the notes column. Figure 2.1 shows typical drawing sheet information.

2.3 Lettering

No particular style of lettering is recommended but the objective is to provide, with reasonable rapidity, distinct uniform letters and figures that will ensure they can be read easily and produce legible copy prints. Where traditional drawing board methods are employed faint guide lines should be used and trainee detailers and engineers should be taught to practise the art of printing which, if neatly executed, increases user confidence. Computer aided detailing software programs provide several lettering systems to create many practical and neat arrangements in the relevant spaces on the drawings.

The minimum font size is 2.5 mm, bearing in mind that microfilming or other reductions in drawing size may be made. Underlining of lettering should not be done except where special emphasis is required. Punctuation marks should not be used unless essential to the sense of the note.

2.4 Dimensions

Arrow heads should have sharp points, touching the lines to which they refer. Dimension lines should be thin but full lines stopped just short of the detail. Dimension figures should be placed immediately above the dimension line and near its centre. The figures should be parallel to the line, arranged so that they can be *read from the bottom or right hand side* of the drawing. Dimensions should normally be given in millimetres and accurate to the nearest whole millimetre.

Steel Detailers' Manual, Third Edition. Alan Hayward and Frank Weare. Third edition revised by Anthony Oakhill.
© 2011 Alan Hayward, Frank Weare and Anthony Oakhill. Published 2011 by Blackwell Publishing Ltd.

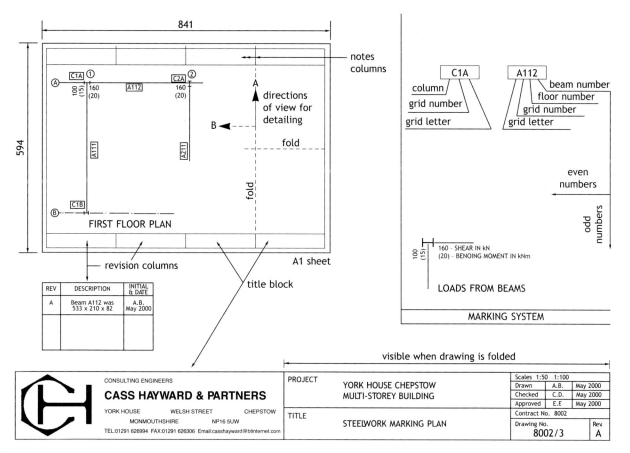

Figure 2.1   Drawing sheets and marking system.

2.5   Projection

Third angle projection should be used whenever possible (see Figure 2.2). With this convention each view is so placed that it represents the side of the object nearest to it in the adjacent view. The notable exception is the base detail on a column, which by convention is shown as in Figure 7.5.

2.6   Scales

Generally scales as follows should be used:

   1 : 5, 1 : 10, 1 : 20, 1 : 25, 1 : 50, 1 : 100, 1 : 200.

Scales should be noted in the title block, and not normally repeated in views. Beams, girders, columns and bracings should preferably be drawn true scale, but may exceptionally be drawn to a smaller longitudinal scale. The section depth and details and other connections must be drawn to scale and in their correct relative positions. A series of sections through a member should be to the same scale, and preferably be arranged in line, in correct sequence.

For bracing systems, lattice girders and trusses a convenient practice is to draw the layout of the centre lines of members to one scale and superimpose details to a larger scale at intersection points and connections.

2.7   Revisions

All revisions must be noted on the drawing in the revision column and every new issue identified by an issue letter, a date and initials of relevant signatories (see Figure 2.1).

2.8   Beam and column detailing conventions

When detailing columns from a floor plan two main views, A viewed from the bottom and B from the right of the plan, must always be given. If necessary, auxiliary views must be added to give the details on the other sides, see Figure 2.2.

Whenever possible columns should be detailed vertically on the drawing, but often it will be more efficient to draw horizontally in which case the base end must be at the right hand side of the drawing with view A at the bottom and view

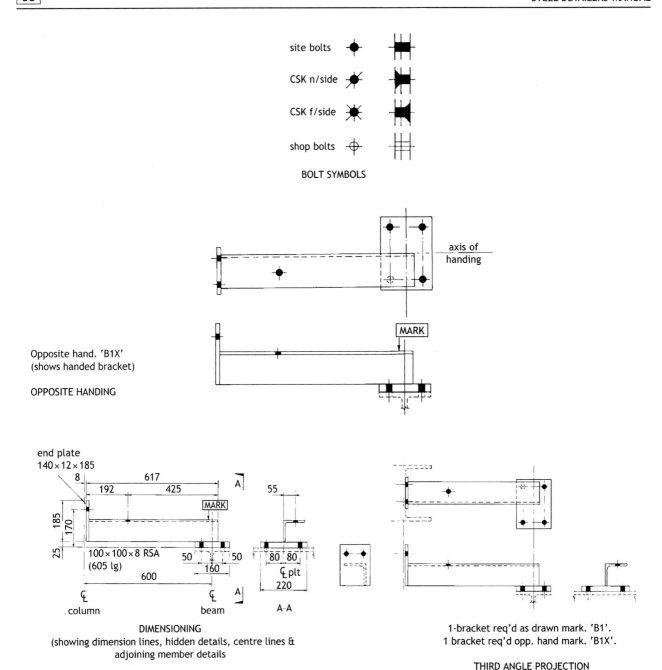

Figure 2.2 Dimensioning and conventions.

B at the top. If columns are detailed vertically the base will naturally be at the bottom with view A on the left of the drawing and view B at the right. Auxiliary views are drawn as necessary. An example of a typical column detail is shown in Figure 7.5.

When detailing a beam from a floor plan, the beam must always be viewed from the bottom or right of the plan. If a beam connects to a seating, end connections must be dimensioned from the bottom flange upwards but if connected by other means (e.g. web cleats, end plates) then end connections must be dimensioned from top flange downwards (see Figure 7.4).

Holes in flanges must be dimensioned from centre-line of web. Rolled steel angles (RSA), channels, etc. should when possible be detailed with the outstanding leg on farside with 'backmark' dimension given to holes.

2.9 Erection marks

An efficient and simple method of marking should be adopted and each loose member or component must have a separate mark. For beam/column structures the allocation of marks for members is shown in Figure 2.2.

On beams the mark should be located on the top flange at the north or east (right-hand) end. On columns the mark should be located on the lower end of the shaft on the flange facing north or east. On vertical bracings the mark should be located at the lower end.

To indicate on a detail drawing where an erection mark is to be painted, the word *mark* contained in a rectangle shall be shown on each detail with an arrow pointing to the position required.

The steelwork contractor normally determines the marking method, taking into account whether or not the mark remains visible after erection. The use of hard stamps is limited due to the possibility of creating notches in highly stressed areas of the steelwork. Similarly, care should be taken when marking weathering steel to ensure it does not damage finish or final appearance.

2.10 Opposite handing

Difficulties frequently arise in both drawing offices and workshops over what is meant by the term *opposite hand*.

Members which are called off on drawings as '1 As Drawn, 1 Opp. Hand' are simply pairs or one right hand and one left hand. A simple illustration of this is the human hand. The left hand is opposite hand to the right hand and vice versa. Any steelwork item must always be opposite handed about a longitudinal centre or datum line and never from end to end. Figure 2.2 shows an example of calling off to opposite hand, with the item referred to also shown to illustrate the principle.

Erection marks are usually placed at the east or north end of an item and opposite handing does not alter this. The erection mark must stay in the position shown on the drawing, i.e. the erection mark is not handed.

2.11 Welds

Welds should be identified using weld symbols as shown in Figure 4.4 and should not normally be drawn in elevation using 'whiskers' or in cross section. In particular cases it may be necessary to draw weld cross sections to an enlarged scale showing butt weld edge preparations such as for complex joints including cruciform type. Usual practice is for workshop butt weld preparations to be shown on separate *weld procedure sheets* not forming part of the drawings. Site welds should be detailed on drawings with the dimensions taking into account allowances for weld shrinkage at site. Space should be allowed around the weld whenever possible so as to allow downhand welding to be used.

2.12 Bolts

Bolts should be indicated using symbolic representation as in Figure 2.2 and should only be drawn with actual bolt and nut where necessary to check particularly tight clearances.

2.13 Holding down bolts

A typical holding down (HD) bolt detail should be drawn out defining length, protrusion above baseplate, thread length, anchorage detail pocket and grouting information and other HD bolts described by notes or schedules. Typical notes are as follows which could be printed onto a drawing or issued separately as a specification.

Notes on holding down bolts

(1) HD bolts shall be cast into foundations using template, accurately to line and level within pockets of size shown to permit tolerance. Immediately after concreting in all bolts shall be 'waggled' to ensure free movement.

(2) Temporary packings used to support and adjust steelwork shall be suitable steel shims placed concentrically with respect to the baseplate. If to be left in place, they shall be positioned such that they are totally enclosed by 30 mm minimum grout cover.

(3) No grouting shall be carried out until a sufficient portion of the structure has been finally adjusted and secured. The spaces to be grouted shall be clear of all debris and free water.

(4) Grout shall have a characteristic strength not less than that of the surrounding concrete nor less than $20\,N/mm^2$. It shall be placed by approved means such that the spaces around HD bolts and beneath the baseplate are completely filled.

(5) Baseplates greater than 400 mm wide shall be provided with at least two grout holes preferably not less than 30 mm in diameter.

(6) Washer plates or other anchorages for securing HD bolts shall be of sufficient size and strength. They shall be designed so that they prevent pull-out failure. The concrete into which HD bolts are anchored shall be reinforced with sufficient overlap and anchorage length so that uplift forces are properly transmitted.

(7) HD bolts shall be of sufficient length to ensure that a minimum of two threads project above the upper nut after tightening.

2.14 Abbreviations

A list of suitable abbreviations for the economic use of space on drawings is given in Table 2.2.

Table 2.2 List of abbreviations.

Description	Abbreviate on drawings
Overall length	O/A
Unless otherwise noted	UON
Diameter	DIA or Φ
Long	LG
Radius	r or RAD
Vertical	VERT
Mark	MK
Dimension	DIM
Near side, far side	N SIDE, F SIDE
Opposite hand	OPP HAND
Centre to centre	C/C
Centre-line	C/L
Horizontal	HORIZ
Drawing	DRG
Not to scale	NTS
Typical	TYP
Nominal	NOM
Reinforced concrete	RC
Floor level	FL
Setting out point	SOP

Description	Abbreviate on drawings
Required	REQD
Section A–A	A–A
Right angle	90°
45 degrees	45°
Slope 1 : 20	(slope symbol 50/1000)
20 number required	20 No
$203 \times 203 \times 52$ kg/m universal column	$203 \times 203 \times 52$ UC
$406 \times 152 \times 60$ kg/m universal beam	$406 \times 152 \times 60$ UB
$150 \times 150 \times 10$ mm angle	$150 \times 150 \times 10$ RSA (or L)
305×102 channel	$305 \times 102\sqsubset$ or 305×102 CHAN
$127 \times 114 \times 29.76$ kg/m joist	$127 \times 114 \times 29.76$ JOIST
$152 \times 152 \times 36$ kg/m structural tee	$152 \times 152 \times 36$ TEE
Girder	GDR
Column	COL
Beam	BEAM
Rolled steel angle	RSA
High strength friction grip bolts	HSFG BOLTS
24 mm diameter bolts grade 8.8	M24 (8.8) BOLTS
Countersunk	CSK
Full penetration butt weld	FPBW
British Standard BS EN 10025: 1993	BS EN 10025: 1993
100 mm length $\times$ 19 diameter shear studs	100×19 SHEAR STUDS
Plate	PLT
Bearing plate	BRG PLT
Packing plate	PACK
Gusset plate	GUSSET
30 mm diameter holding down bolts grade 8.8, 600 mm long	M30 (8.8) HD BOLTS 600 LG
Flange plate	FLG
Web plate	WEB
Intermediate stiffener	STIFF
Bearing stiffener	BRG STIFF
Fillet weld	FW (but use welding symbols!)
Machined surface	m/c ↙
Fitted to bear	FIT
Cleat	CLEAT
35 pitches at 300 centres = 10 500	35×300 c/c = 10 500
70 mm wide $\times$ 12 mm thick plate	70×12 PLT
120 mm wide $\times$ 10 mm thick $\times$ 300 mm long plate	120×10 PLT $\times$ 300
25 mm thick	25 THK
80 mm $\times$ 80 mm plate $\times$ 6 mm thick	80 SQ $\times$ 6 PLT

3 Design Guidance

3.1 General

Limited design guidance is included in this manual for selecting *simple connections* and *simple baseplates* which can be carried out by the detailer without demanding particular skills. Other connections including *moment connections* and the design of members such as beams, girders, columns, bracings and lattice structures will require specific design calculations. Load capacities for members are contained in the Design Guide to BS 5950[5] and from other literature as given in the Further Reading.

Capacities of bolts and welds to BS 5950 are included in Tables 3.5, 3.6 and 3.7 so that detailers can proportion elementary connections such as welds and bolts to gusset plates etc.

3.2 Load capacities of simple connections

Ultimate load capacities for a range of simple web angle cleat/end plate type beam/column and beam/beam connections for universal beams are given in Tables 3.1, 3.2 and 3.3. The capacities must be compared with *factored* loads to BS 5950. The tables indicate whether bolt shear, bolt bearing, web shear or weld strength are critical so that

different options can be examined. The range of coverage is listed at the foot of this page.

Capacities in kN are presented under the following symbols:

Connection to beam	Bc—RSA cleats
	Be—End plates
Connection to column	S1—one-sided connection—maximum
	S2—two-sided connection—total reaction from two incoming beams sharing the same bolt group

Worked example
The following example illustrates use of Tables 3.1, 3.2 and 3.3.

Question
A beam of size 686 × 254 × 140 UB in grade S275 steel has a factored end reaction of 750 kN. Design the connection using RSA web cleats:

(a) *to a perimeter column size 305 × 305 × 97 UC, of grade S275 steel via its flange*
(b) *to a similar internal column via its web, forming a two sided connection with another beam having the same reaction.*

Table	Steel grade	M20 bolts grade	Grade S275 RSA web cleats		Grade S275 end plates		Number of bolt rows to column/beam	Welds to end plate	
			To column	To beam	To columns	To beams		N11 to N6	N5 to N1
3.1	S275	4.6	100 × 100 × 10	90 × 90 × 10	200 × 10	160 × 8	Range	8 mm	6 mm
3.2	S275	8.8	100 × 100 × 10	90 × 90 × 10	200 × 10	160 × 10	N11 to N1	fillet	fillet
3.3	S355	8.8	100 × 100 × 10	90 × 90 × 10	200 × 10	160 × 10		welds	welds

Steel Detailers' Manual, Third Edition. Alan Hayward and Frank Weare. Third edition revised by Anthony Oakhill.
© 2011 Alan Hayward, Frank Weare and Anthony Oakhill. Published 2011 by Blackwell Publishing Ltd.

Answer

(a) To perimeter columns

Connection to beam:

$686 \times 254 \times 140$ UB – web thickness 12.4 mm

From Table 3.1 (grade 4.6 bolts) maximum value of Bc = 556 kN for N8 type, which is insufficient. Capacity cannot be increased by thicker webbed beam because bolt shear governs (because value is not in italics).

So try grade 8.8 bolts:

From Table 3.2 value of Bc = 770 kN for 12 mm web
for N8 type
= 898 kN for 14 mm web
for N8 type

Interpolation for 12.4 mm web gives

Bc = 796 kN > 750 ACCEPT

Connection to column: $305 \times 305 \times 97$ UC – flange thickness 15.4 mm

From Table 3.2 value of S1 = 1449 kN for 14 mm flange
= 1449 kN for 16 mm flange
Therefore S1 = 1449 kN for 15.4 mm
flange > 750 ACCEPT

Therefore connection is N8 with $100 \times 100 \times 10$ RSA cleats, i.e. 8 rows of M20 (8.8) bolts.

(b) To internal column connection to beam

Connection to beam

As for (a) i.e. N8 type using grade 8.8 bolts.

Connection to column: $305 \times 305 \times 97$ UC – web thickness 9.9 mm.

From Table 3.2, value of S2 = 1178 kN for 8 mm web
= 1472 kN for 10 mm web

Interpolation for 9.9 mm web gives S2 = 1457 kN < 2 × 750 = 1500 kN

Therefore insufficient, but note that bolt bearing is critical (because value is in italics) so try grade S355 steel for column.

From Table 3.3, value of S2 = 1408 kN for 8 mm web
= 1760 kN for 10 mm web

Interpolation for 9.9 mm web gives

S2 = 1742 kN > 1500 ACCEPT

Alternatively try larger diameter bolts:

For M22 (8.8) bolt:

From Table 3.5: giving capacities of single bolts: double shear value = 227 kN

bearing to 2/10 mm S275 cleats 2 × 101 = 202 kN

bearing to UB web S275

9 mm thick 91 kN

10 mm thick 101 kN

Interpolation for 9.9 mm thick gives 100 kN

Therefore bearing to UC web governs.

So capacity is 16 bolts × 100 = 1600 kN > 1500 ACCEPT

Therefore M22 (8.8) bolts can be used instead of using grade S355 steel for the column.

3.3 Sizes and load capacity of simple column bases

Ultimate capacities and baseplate thicknesses using grade S275 steel for a range of simple square column bases with universal column or square hollow section columns are given in Table 3.4. These capacities must be compared with factored loads to BS 5950.

Baseplate thickness is derived to BS 5950-1 clause 4.13.2.2:

$$tp = c\left[\frac{3w}{pyp}\right]^{\frac{1}{2}}$$

where

c   is the largest perpendicular distance from the edge of the effective portion of the baseplate to the face of the column cross-section

pyp is the design strength of the baseplate

w   is the pressure under the baseplate, based on an assumed uniform distribution of pressure throughout the effective portion.

Worked example

Question

The following example illustrates use of Table 3.4. A $305 \times 305 \times 97$ UC column carries a factored vertical load of 3000 kN at the base. The foundation concrete has an ultimate strength of 30 N/mm². Select a baseplate size.

Answer

From Table 3.4 width of base for concrete strength 30 N/mm² is 500 mm for P = 300 kN.

Thickness = 30 mm

Therefore baseplate minimum size is $500 \times 30 \times 500$ in grade S275 steel.

Table 3.1 Simple connections, bolts grade 4.6, members grade S275. See Figure 3.1.

Type	UB sizes for beam	Symbols	\[Thickness (mm) of beam web or column web/flange connected\] 4	6	8	10	12	14	16	18	20
N11	914×419, 914×305	Bc / S1	– / –	– / –	– / –	– / 862	– / 862	803 / 862	803 / 862	803 / 862	803 / 862
		Be / S2	– / –	– / –	– / –	– / *1531*	– / 1725	*1714* / 1725	*1810* / 1725	*1810* / 1725	*1803* / 1725
N10	914×419, 838×292	Bc / S1	– / –	– / –	– / –	– / 784	– / 784	721 / 784	721 / 784	721 / 784	721 / 784
		Be / S2	– / –	– / –	– / *1392*	– / 1568	– / 1568	*1558* / 1568	*1642* / 1568	*1642* / 1568	*1642* / 1568
N9	914×419, 838×292, 762×267	Bc / S1	– / –	– / –	– / –	– / 706	638 / 706	638 / 706	638 / 706	638 / 706	638 / 706
		Be / S2	– / –	– / –	– / *1253*	– / 1411	*1202* / 1411	*1474* / 1411	*1474* / 1411	*1474* / 1411	*1474* / 1411
N8	914×419, 838×292, 762×267, 686×254	Bc / S1	– / –	– / –	– / 627	556 / 627	556 / 627	556 / 627	556 / 627	556 / 627	556 / 627
		Be / S2	– / –	– / *835*	– / *1114*	*890* / 1254	*1068* / 1254	*1247* / 1254	*1306* / 1254	*1306* / 1254	*1306* / 1254
N7	838×292, 762×267, 686×254, 610×305	Bc / S1	– / –	– / 549	– / 549	473 / 549	473 / 549	473 / 549	473 / 549	473 / 549	473 / 549
		Be / S2	– / –	– / *731*	– / *974*	*779* / 1098	*935* / 1098	*1091* / 1098	*1138* / 1098	*1138* / 1098	*1138* / 1098
N6	686×254, 610×305, 533×210	Bc / S1	– / –	– / 470	– / 470	390 / 470	390 / 470	390 / 470	390 / 470	390 / 470	390 / 470
		Be / S2	– / –	*339* / *626*	*554* / *835*	*668* / 941	*801* / 941	*935* / 941	*970* / 941	*970* / 941	*970* / 941
N5	457×191, 406×178	Bc / S1	*139* / *348*	*200* / 392	*267* / 392	314 / 392	314 / 392	– / 392	– / 392	– / 392	– / 392
		Be / S2	– / *348*	*347* / *522*	*462* / *696*	*557* / 784	*610* / 784	– / 784	– / 784	– / 784	– / 784
N4	457×152, 406×178, 356×171	Bc / S1	*98* / *278*	*147* / 314	*196* / 314	228 / 314	228 / 314	– / 314	– / 314	– / 314	– / 314
		Be / S2	– / *278*	*277* / *418*	*370* / *557*	*445* / 627	*484* / 627	– / 627	– / 627	– / 627	– / 627
N3	406×140, 356×127, 305×127, 254×102	Bc / S1	*65* / *209*	*97* / 235	*129* / 235	152 / 235	– / 235	– / 235	– / 235	– / 235	– / 235
		Be / S2	*139* / *209*	*208* / *313*	*277* / *418*	*347* / 470	– / 470	– / 470	– / 470	– / 470	– / 470
N2	305×127, 305×102, 254×102, 203×102	Bc / S1	*35* / *139*	*53* / 157	*70* / 157	84 / 157	– / 157	– / 157	– / 157	– / 157	– / 157
		Be / S2	*92* / *139*	*139* / *209*	*185* / *278*	*231* / 314	– / 314	– / 314	– / 314	– / 314	– / 314
N1	254×146, 203×133	Bc / S1	*35* / *70*	*53* / 78	*70* / 78	84 / 78	– / 78	– / 78	– / 78	– / 78	– / 78
		Be / S2	*92* / *70*	*139* / *104*	*185* / *139*	*231* / 157	– / 157	– / 157	– / 157	– / 157	– / 157

Connection to beam:
Bc – RSA cleats
Be – End plates

Connection to column:
S1 – RSA cleats or end plates
S2 – RSA cleats or end plates

Bc – RSA cleats – capacity in kN – lesser of bolt shear and bolt bearing to web. Value in italics is bolt bearing where less.
Be – End plates – capacity in kN – bolt bearing to web. Value in italics is bolt bearing to web.
S1 – RSA cleats or end plates – capacity in kN – lesser of web shear or weld strength. Value in italics is weld strength where less.
S2 – RSA cleats or end plates – capacity in kN – Least of bolt shear, bolt bearing to web, or bolt bearing to cleat. Value in italics is bolt bearing where the least.

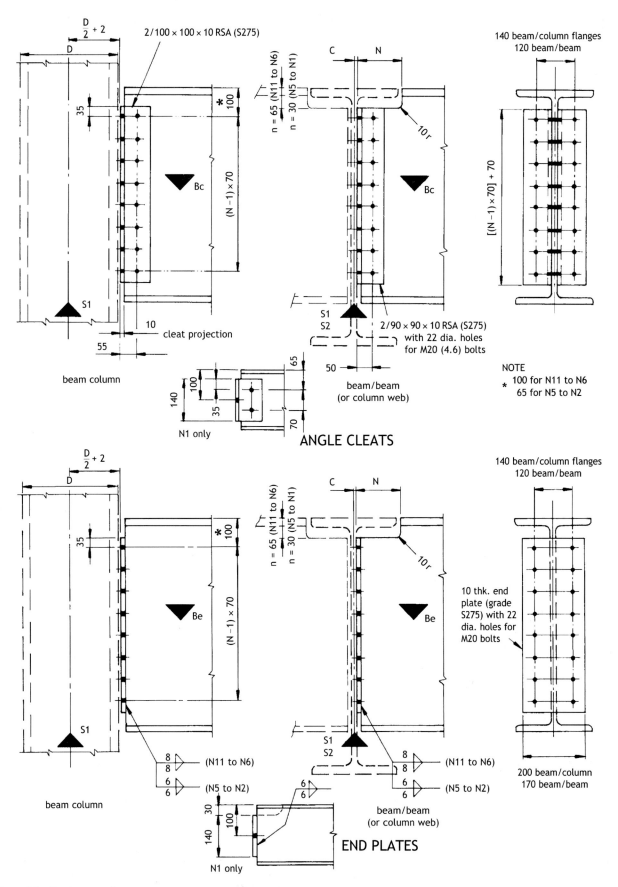

Figure 3.1 Simple connections.

Table 3.2　Simple connections, bolts grade 8.8, members grade S275. See Figure 3.1.

Thickness (mm) of beam web or column web/flange connected

Type	UB sizes for beam	Symbols	4	6	8	10	12	14	16	18	20
N11	914 × 419	Be S2	– –	– –	– **1619**	– **2024**	– **2429**	*1714* **2834**	*1810* **3238**	*1810* **3643**	*1810* **4002**
		Bc S1	– –	– –	– **1619**	– **2001**	– **2001**	*1304* **2001**	*1490* **2001**	*1676* **2001**	*1862* **2001**
N10	914 × 419, 838 × 292	Be S2	– –	– –	– **1472**	– **1840**	– **2208**	*1558* **2576**	*1642* **2944**	*1642* **3312**	*1642* **3634**
		Bc S1	– –	– –	– **1472**	– **1817**	– **1817**	*1169* **1817**	*1336* **1817**	*1503* **1817**	*1670* **1817**
N9	914 × 419, 838 × 292, 762 × 267	Be S2	– –	– –	– **1325**	– **1656**	*1202* **1987**	*1474* **2318**	*1474* **2650**	*1474* **2980**	*1474* **3266**
		Bc S1	– –	– –	– **1325**	– **1633**	*886* **1633**	*1034* **1633**	*1182* **1633**	*1329* **1633**	*1477* **1633**
N8	914 × 419, 838 × 292, 762 × 267, 686 × 254	Be S2	– –	– **883**	– **1178**	*890* **1472**	*1068* **1766**	*1247* **2061**	*1306* **2355**	*1306* **2650**	*1306* **2898**
		Bc S1	– –	– **883**	– **1178**	*642* **1449**	*770* **1449**	*898* **1449**	*1027* **1449**	*1155* **1449**	*1283* **1449**
N7	838 × 292, 762 × 267, 686 × 254, 610 × 305, 610 × 229	Be S2	– –	– **773**	– **1030**	*779* **1288**	*935* **1546**	*1091* **1803**	*1138* **2061**	*1138* **2318**	*1138* **2530**
		Bc S1	– –	– **773**	– **1030**	*545* **1265**	*654* **1265**	*763* **1265**	*872* **1265**	*981* **1265**	*1090* **1265**
N6	686 × 254, 610 × 305, 610 × 229, 533 × 210	Be S2	– –	– **663**	*554* **883**	*668* **1104**	*801* **1325**	*935* **1546**	*970* **1766**	*970* **1987**	*970* **2162**
		Bc S1	– –	– **663**	*359* **883**	*448* **1081**	*457* **1081**	*628* **1081**	*717* **1081**	*807* **1081**	*897* **1081**
N5	457 × 191, 406 × 178	Be S2	– **368**	*347* **552**	*462* **736**	*557* **920**	*610* **1104**	– **1288**	– **1472**	– **1656**	– **1794**
		Bc S1	– **368**	*212* **552**	*282* **736**	*353* **897**	*423* **697**	– **897**	– **897**	– **897**	– **897**
N4	457 × 191, 406 × 178, 356 × 171	Be S2	*185* **294**	*277* **442**	*370* **589**	*445* **736**	*484* **883**	– **1030**	– **1178**	– **1325**	– **1426**
		Bc S1	*104* **294**	*156* **442**	*208* **589**	*260* **713**	*311* **713**	– **713**	– **713**	– **713**	– **713**
N3	406 × 178, 356 × 171, 305 × 165, 254 × 146	Be S2	*139* **221**	*208* **331**	*277* **442**	*347* **552**	– **662**	– **773**	– **883**	– **994**	– **1058**
		Bc S1	*68* **221**	*103* **331**	*137* **442**	*171* **529**	– **529**	– **529**	– **529**	– **529**	– **529**
N2	305 × 165, 254 × 146, 203 × 133	Be S2	*92* **147**	*139* **221**	*185* **294**	*231* **368**	– **442**	– **515**	– **589**	– **662**	– **690**
		Bc S1	*37* **147**	*56* **221**	*74* **294**	*93* **345**	– **345**	– **345**	– **345**	– **345**	– **345**
N1	254 × 146, 203 × 133	Be S2	*92* **74**	*139* **110**	*185* **147**	*231* **184**	– **221**	– **258**	– **294**	– **331**	– **322**
		Bc S1	*37* **74**	*56* **110**	*74* **147**	*93* **161**	– **161**	– **161**	– **161**	– **161**	– **161**

Value in italics is bolt bearing where the least. Value shown in bold.

Connection to beam:
Bc – RSA cleats – capacity in kN – lesser of bolt shear and bolt bearing to web. Value in italics is bolt bearing where less.
Be – End plates – capacity in kN – lesser of web shear or weld strength. Value in italics is weld strength where less.

Connection to column:
S1 – RSA cleats or end plates – capacity in kN　Least of bolt shear, bolt bearing to web, or bolt bearing to cleat. Value in italics is bolt bearing where least.
S2 – RSA cleats or end plates – capacity in kN　type if bolt bearing to cleats is least.

Table 3.3 Simple connections, bolts grade 8.8, members grade S355. See Figure 3.1.

Values in each cell are given as *italic* **bold** (italic = bolt-bearing value where less; bold = governing capacity).

Type	UB sizes for beam	Symbol	4	6	8	10	12	14	16	18	20
N11	914 × 419	Bc / Be	–	–	**1935**	**2001**	**2001**	*1558* **2001**	*1781* **2001**	*1862* **2001**	*1862* **2001**
		S1 / S2	–	–	**1935**	**2420**	**2904**	*1810* **3388**	*1810* **3872**	*1810* **4002**	*1810* **4002**
N10	914 × 419; 838 × 292	Bc / Be	–	–	**1760**	**1817**	**1817**	*1398* **1817**	*1597* **1817**	*1670* **1817**	*1670* **1817**
		S1 / S2	–	–	**1760**	**2200**	**2640**	*1642* **3080**	*1642* **3520**	*1642* **3634**	*1642* **3634**
N9	914 × 419; 838 × 292; 762 × 267	Bc / Be	–	–	**1584**	**1633**	*1060* **1633**	*1236* **1633**	*1413* **1633**	*1477* **1633**	*1477* **1633**
		S1 / S2	–	–	**1584**	**1980**	*1474* **2376**	*1474* **2772**	*1474* **3168**	*1474* **3266**	*1474* **3266**
N8	914 × 419; 838 × 292; 762 × 267; 686 × 254	Bc / Be	–	**1056**	**1408**	*768* **1449**	*921* **1449**	*1074* **1449**	*1228* **1449**	*1283* **1449**	*1283* **1449**
		S1 / S2	–	**1056**	**1408**	**1760**	*1306* **2112**	*1306* **2464**	*1306* **2816**	*1306* **2898**	*1306* **2898**
N7	838 × 292; 762 × 267; 686 × 254; 610 × 305; 610 × 229	Bc / Be	–	**924**	**1231**	*652* **1265**	*782* **1265**	*912* **1265**	*1043* **1265**	*1090* **1265**	*1090* **1265**
		S1 / S2	–	**924**	**1231**	*1000* **1540**	*1138* **1848**	*1138* **2156**	*1138* **2464**	*1138* **2530**	*1138* **2530**
N6	686 × 254; 610 × 305; 533 × 210; 610 × 229	Bc / Be	–	**792**	*429* **1056**	*536* **1081**	*643* **1081**	*751* **1081**	*858* **1081**	*897* **1081**	*897* **1081**
		S1 / S2	–	**792**	*716* **1056**	*857* **1320**	*970* **1584**	*970* **1848**	*970* **2112**	*970* **2162**	*970* **2162**
N5	457 × 191; 406 × 178	Bc / Be	**440**	*253* **660**	*337* **880**	*422* **897**	*506* **897**	**897**	**897**	**897**	**897**
		S1 / S2	**440**	*447* **660**	*596* **880**	*610* **1100**	*610* **1320**	**1540**	**1760**	**1794**	**1794**
N4	457 × 191; 406 × 178; 356 × 171; 356 × 127	Bc / Be	*124* **352**	*186* **528**	*248* **704**	*310* **713**	*372* **713**	**713**	**713**	**713**	**713**
		S1 / S2	*239* **352**	*358* **528**	*477* **704**	*484* **880**	*484* **1056**	**1232**	**1408**	**1426**	**1426**
N3	406 × 178; 356 × 171; 305 × 165; 254 × 146; 356 × 127; 305 × 127	Bc / Be	*82* **264**	*123* **396**	*164* **528**	*205* **529**	**529**	**529**	**529**	**529**	**529**
		S1 / S2	*179* **264**	*268* **396**	*358* **528**	*358* **660**	*358* **792**	*358* **924**	*529* **1056**	*529* **1058**	*529* **1058**
N2	305 × 165; 254 × 146; 203 × 133; 305 × 102; 254 × 102; 203 × 102	Bc / Be	*44* **176**	*66* **264**	*89* **345**	*111* **345**	**345**	**345**	**345**	**345**	**345**
		S1 / S2	*119* **176**	*179* **264**	*231* **352**	*231* **440**	**528**	**666**	**690**	**690**	**690**
N1	254 × 146; 203 × 133	Bc / Be	*44* **88**	*66* **132**	*89* **176**	*111* **220**	**264**	**308**	**322**	*161* **322**	*161* **322**
		S1 / S2	*44* **88**	*66* **132**	*89* **176**	*111* **220**	**264**	**308**	**322**	*161* **322**	*161* **322**

Header span: Thickness (mm) of beam web or column web/flange connected (columns 4–20).

Connection to beam:
Bc – RSA cleats – capacity in kN — lesser of bolt shear and bolt bearing to web. Value in italics is bolt bearing where less. Value shown in bold.
Be – End plates – capacity in kN — lesser of web shear or bolt bearing. Value in italics is weld strength where less.

Connection to column:
S1 – RSA cleats or end plates – capacity in kN — Least of bolt shear, bolt bearing to web, or bolt bearing to cleat. Value in italics is bolt bearing to cleat. Value shown in bold is bolt bearing where the least.
S2 – RSA cleats or end plates – capacity in kN — capacity in kN type if bolt bearing to cleats is least.

Table 3.4 Simple column bases (see Figure 3.2 on page 48).

WIDTH OF BASE B (mm)

Concrete Strength f_{cu} = 30 N/mm²

UC	SHS	200	225	250	275	300	325	350	375	400	425	450	475	500	525	550	575	600	625	650	675	700	725	750
P (kN)	100	480	600	750	900	1080	1260	1470	1680	1920	2160	2430	2700	3000	3300	3630	3960	4320	4680	5070	5460	5880	6300	6750
152 × 152	150	15	20	25	25	30	35	–	–	–	–	–	–	–	–	–	–	–	–	–	–	–	–	–
203 × 203	200	–	–	15	20	25	25	30	35	40	40	45	50	55	50	55	55	60	–	–	–	–	–	–
254 × 254	250	–	–	–	–	15	20	25	25	30	35	40	40	45	40	45	50	55	55	60	65	–	–	–
305 × 305	300	–	–	–	–	–	–	15	15	25	25	30	30	40	35	40	40	45	50	55	55	60	65	70
356 × 368	350	–	–	–	–	–	–	–	–	–	25	25	25	30	25	30	35	40	40	45	45	55	55	60
356 × 406	400	–	–	–	–	–	–	–	–	–	20	15	20	25	20	25	25	30	35	40	40	45	50	55

Concrete Strength f_{cu} = 40 N/mm²

UC	RHS	200	225	250	275	300	325	350	375	400	425	450	475	500	525	550	575	600	625	650	675	700	725	750
P (kN)	100	640	810	1000	1210	1440	1690	1960	2250	2560	2890	3240	3610	4000	4410	4840	5290	5760	6250	6750	7290	7840	8410	9000
152 × 152	150	20	25	25	30	35	30	35	40	45	50	55	50	55	55	60	–	–	–	–	–	–	–	–
203 × 203	200	–	–	20	25	25	25	25	30	35	40	45	50	55	50	55	55	60	65	–	–	–	–	–
254 × 254	250	–	–	–	–	20	25	20	25	25	30	35	30	35	40	45	50	55	55	65	75	70	75	85
305 × 305	300	–	–	–	–	–	–	–	25	20	25	25	25	25	30	35	40	45	50	55	65	60	65	80
356 × 368	350	–	–	–	–	–	–	–	–	–	–	20	25	20	25	30	35	35	40	50	55	55	55	70
356 × 406	400	–	–	–	–	–	–	–	–	–	–	–	–	20	25	25	25	35	40	45	50	55	55	70

Concrete Strength f_{cu} = 50 N/mm²

UC	RHS	200	225	250	275	300	325	350	375	400	425	450	475	500	525	550	575	600	625	650	675	700	725	750
P (kN)	100	800	1010	1250	1510	1800	2110	2450	2810	3200	3610	4050	4510	5000	5510	6050	6610	7200	7810	8450	9110	9800	10510	11250
152 × 152	150	20	25	25	30	35	40	45	50	55	50	55	55	60	55	60	65	70	75	75	75	80	85	–
203 × 203	200	–	–	–	25	30	35	40	45	50	55	50	55	55	60	65	65	70	75	80	75	80	85	–
254 × 254	250	–	–	–	–	20	25	30	35	40	45	50	55	60	65	65	70	75	80	85	–	–	–	–
305 × 305	300	–	–	–	–	–	25	30	35	40	45	50	55	60	65	70	75	80	85	–	–	–	–	–
356 × 368	350	–	–	–	–	–	–	20	25	30	35	40	45	50	55	60	65	70	75	80	85	–	–	–
356 × 406	400	–	–	–	–	–	–	–	–	20	25	30	35	40	45	50	55	60	65	70	75	80	85	–

The details are suitable for column bases comprising mainly vertical load only. The holding down bolts shown are suitable for nominal transverse or uplift forces and for securing/adjusting the steelwork during erection. A typical length is 24 diameter. Fixed bases (i.e. carrying bending moments) require larger HD bolts and increase in size of welds. Length of HD bolts must be adequate for anchorage and sufficient lapping reinforcement in the base supplied. Gusseted bases may be required in carrying loads in excess of those tabled or where bending moments are very significant. They are costly and simple column bases as shown should be used whenever possible.

P(kN). Ultimate vertical load capacity of concrete beneath base (stress 0.6 f_{cu}) to BS 5950-1:2000.

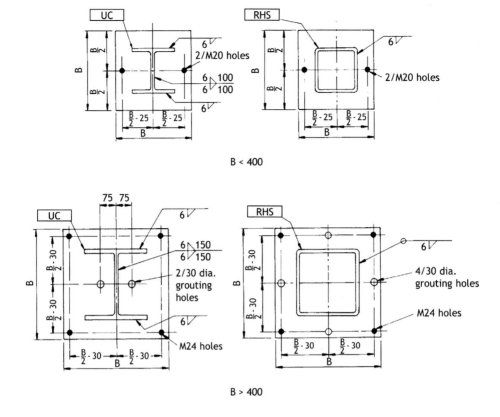

B < 400

B > 400

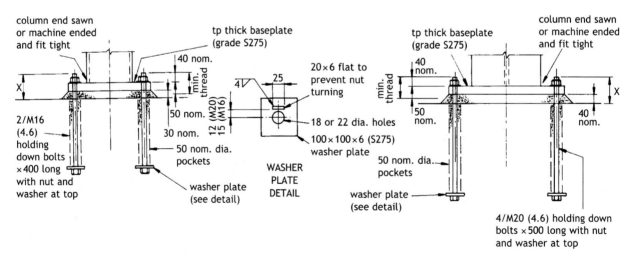

Figure 3.2   Simple column bases.

Table 3.5 Black bolt capacities.

4.6 Bolts in material grades S275 and S355

Diam of Bolt mm	Tensile Stress Area mm²	Tensile Cap kN	Shear Value		Bearing Value of bolt at 435N/mm² and end distance equal to 2 x bolt diameter Thickness in mm of Plate Passed Through										
			Single Shear kN	Double Shear kN	5	6	7	8	9	10	12.5	15	20	25	30
12	84.3	16.4	13.5	27.0	26	31	0	0	0	0	0	0	0	0	0
16	157	30.6	25.1	50.2	34	41	48	55	0	0	0	0	0	0	0
20	245	47.8	39.2	78.4	43	52	60	69	78	87	0	0	0	0	0
22	303	59.1	48.5	97.0	47	57	67	76	86	95	120	0	0	0	0
24	353	68.8	56.5	113	52	62	73	83	94	104	131	0	0	0	0
27	459	89.5	73.4	147	58	70	82	94	106	117	147	176	0	0	0
30	561	109	89.8	180	65	78	91	104	117	131	163	196	0	0	0

8.8 Bolts in material grade S275

Diam of Bolt mm	Tensile Stress Area mm²	Tensile Cap kN	Shear Value		Bearing Value of plate at 460N/mm² and end distance equal to 2 x bolt diameter Thickness in mm of Plate Passed Through										
			Single Shear kN	Double Shear kN	5	6	7	8	9	10	12.5	15	20	25	30
12	84.3	37.9	31.6	63.2	27	33	38	44	49	55	69	0	0	0	0
16	157	70.7	58.9	118	36	44	51	58	66	73	93	110	147	0	0
20	245	110	91.9	184	46	55	64	73	82	92	115	138	184	0	0
22	303	136	114	227	50	60	70	81	91	101	127	152	202	253	0
24	353	159	132	265	55	66	77	88	99	110	138	166	221	276	0
27	459	207	172	344	62	74	86	99	112	124	155	186	248	310	373
30	561	252	210	421	69	82	96	110	124	138	173	207	276	345	414

8.8 Bolts in material grade S355

Diam of Bolt mm	Tensile Stress Area mm²	Tensile Cap kN	Shear Value		Bearing Value of plate at 550N/mm² and end distance equal to 2 x bolt diameter Thickness in mm of Plate Passed Through										
			Single Shear kN	Double Shear kN	5	6	7	8	9	10	12.5	15	20	25	30
12	84.3	37.9	31.6	63.2	33	39	46	52	59	66	0	0	0	0	0
16	157	70.7	58.9	118	44	52	61	70	79	88	110	132	0	0	0
20	245	110	91.9	184	55	66	77	88	99	110	138	165	220	0	0
22	303	136	114	227	60	72	84	96	109	121	151	181	242	0	0
24	353	159	132	265	66	79	92	106	119	132	165	198	264	330	0
27	459	207	172	344	74	89	104	119	134	148	186	223	297	371	0
30	561	252	210	421	82	99	116	132	148	165	206	247	330	413	495

Values printed in bold type are less than the single shear value of the bolt. Values printed in ordinary type are greater than the single shear value and less than the double shear value. Values printed in italic type are greater than the double shear value.
Bearing values are governed by the strength of the bolt

Table 3.6   HSFG bolt capacities.

In material grade S275

Diam of Bolt mm	Proof Load Of Bolt kN	Tensile Cap kN	Slip Value Single Shear kN	Slip Value Double Shear kN	Bearing Value of plate at 825N/mm² and end distance equal to 3 x bolt diameter — Thickness in mm of Plate Passed Through										
					5	6	7	8	9	10	12.5	15	20	25	30
12	49.4	44.5	24.5	48.9	49	0	0	0	0	0	0	0	0	0	0
16	92.1	82.9	45.6	91.2	66	79	92	0	0	0	0	0	0	0	0
20	144	130	71.3	143	82	99	116	132	148	0	0	0	0	0	0
22	177	159	87.6	175	90	109	127	145	163	181	0	0	0	0	0
24	207	186	102	205	99	119	139	158	178	198	247	0	0	0	0
27	234	211	116	232	111	134	156	178	200	223	278	0	0	0	0
30	286	257	142	283	124	148	173	198	233	247	309	0	0	0	0

In material grade S355

Diam of Bolt mm	Proof Load Of Bolt kN	Tensile Cap kN	Slip Value Single Shear kN	Slip Value Double Shear kN	Bearing Value of plate at 1065N/mm² and end distance equal to 3 x bolt diameter — Thickness in mm of Plate Passed Through										
					5	6	7	8	9	10	12.5	15	20	25	30
12	49.4	44.5	24.5	48.9	63	0	0	0	0	0	0	0	0	0	0
16	92.1	82.9	45.6	91.2	85	102	0	0	0	0	0	0	0	0	0
20	144	130	71.3	143	106	128	149	0	0	0	0	0	0	0	0
22	177	159	87.6	175	117	141	164	187	0	0	0	0	0	0	0
24	207	186	102	205	128	153	179	204	230	0	0	0	0	0	0
27	234	211	116	232	144	173	201	230	259	0	0	0	0	0	0
30	286	257	142	283	160	192	224	256	288	0	0	0	0	0	0

Values printed in bold type are less than the single shear value of the bolt. Values printed in ordinary type are greater than the single shear value and less than the double shear value. Values printed in italic type are greater than the double shear value.
Bearing values are governed by the strength of the plate
Slip Capacity based on a slip factor of 0.45

Table 3.7 Weld capacities.

(a) Strength of fillet welds.

Leg length mm	Throat thickness mm	Capacity at 215 N/mm^2 kN/m	Leg length mm	Throat thickness mm	Capacity at 215 N/mm^2 kN/m
3.0	2.12	456	12.0	8.49	1824
4.0	2.83	608	15.0	10.61	2280
5.0	3.54	760	18.0	12.73	2737
6.0	4.24	912	20.0	14.14	3041
8.0	5.66	1216	22.0	15.56	3345
10.0	7.07	1520	25.0	17.68	3801

Capacities with grade E43 electrodes to BS EN 499 and BS EN 22553. Grades of steel S275 and S355.

(b) Strength of full penetration butt welds.

Thickness mm	Shear at $0.6 \times Py$ kN/m	Tension or compression at Py kN/m	Thickness mm	Shear at $0.6 \times Py$ kN/m	Tension or compression at Py kN/m
Grade of steel S275					
6.0	990	1650	22.0	3498	5830
8.0	1320	2200	25.0	3975	6625
10.0	1650	2750	28.0	4452	7420
12.0	1980	3300	30.0	4770	7950
15.0	2475	4125	35.0	5565	9275
18.0	2862	4770	40.0	6360	10600
20.0	3180	5300	45.0	6885	11475
Grade of steel S355					
6.0	1278	2130	22.0	4554	7590
8.0	1704	2840	25.0	5175	8625
10.0	2130	3550	28.0	5796	9660
12.0	2556	4260	30.0	6210	10350
15.0	3195	5325	35.0	7245	12075
18.0	3726	6210	40.0	8280	13800
20.0	4140	6900	45.0	9180	15300

4 Detailing Data

The following data provide useful information for the detailing of steelwork. The dimensional information on standard sections is given by permission of Tata Steel (previously Corus). These sections are widely used in many other countries.

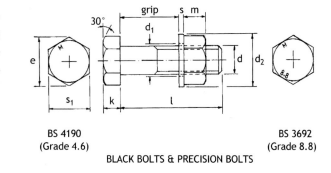

BS 4190
(Grade 4.6)

BS 3692
(Grade 8.8)

BLACK BOLTS & PRECISION BOLTS

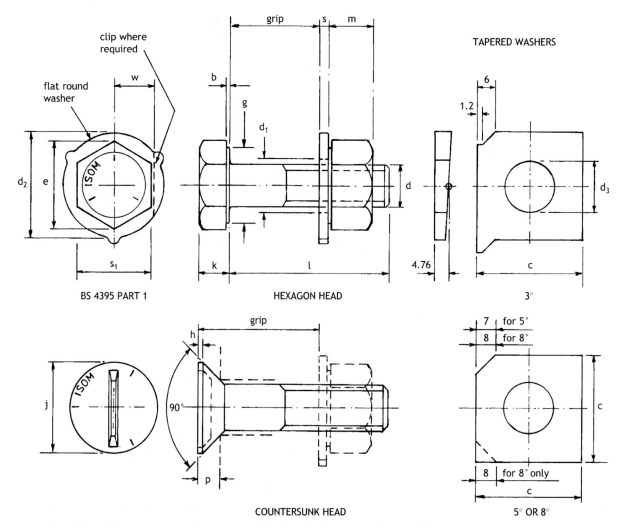

BS 4395 PART 1

HEXAGON HEAD

TAPERED WASHERS

3°

COUNTERSUNK HEAD

5° OR 8°

Steel Detailers' Manual, Third Edition. Alan Hayward and Frank Weare. Third edition revised by Anthony Oakhill.
© 2011 Alan Hayward, Frank Weare and Anthony Oakhill. Published 2011 by Blackwell Publishing Ltd.

Table 4.1 Dimensions of black bolts.

Dimensions to BS 4190 (nearest mm)

Bolts and Nuts				Bolts				Nuts		Washers (to BS 4320)			
Nominal Size	Pitch of Thread	Width Across Flats (max)	Width Across Corners (max)	Depth of Head (max)	Standard Length of Thread l			Depth m (max)		Diameter			Thickness (Nom)
										Inside (Nom)	Outside (max) d2		
d		S1	e	k	≤125	<200	>200	Std	Thin	d1	Normal	Large	s
(M12)	1.74	19	22	9	30	36	49	11	7	14	24	28	3
M16	2.0	24	28	11	38	44	57	14	9	18	30	34	3
M20	2.5	30	35	14	46	52	65	17	9	22	37	39	3
(M22)	2.5	32	37	15	50	56	69	19	10	24	39	44	3
M24	3.0	36	42	16	54	60	73	20	10	26	44	50	4
(M27)	3.0	41	47	18	60	66	79	23	12	30	50	56	4
M30	3.5	46	53	20	66	72	85	25	12	33	56	60	4
(M33)	3.5	50	58	22	72	78	91	27	14	36	60	66	5
M36	4.0	55	64	24	78	84	97	30	14	39	66	76	5

Mechanical Properties

Nominal Size	Tensile Stress Area	Grade 4.6 (BS 4190)		Grade 8.8 (BS 3692)	
		Ultimate Load	Proof Load	Ultimate Load	Proof Load
d	mm²	kN	kN	kN	kN
(M12)	84.3	33.1	18.7	66.2	48.1
M16	157	61.6	34.8	123	89.6
M20	245	96.1	54.3	192	140
(M22)	303	118.8	67.3	238	173
M24	353	138	78.2	277	201
(M27)	459	180	102	360	262
M30	561	220	124	439	321
(M33)	694	272	154	544	396
M36	817	321	181	641	466

Recommended Bolt and Nut Combinations

Grade of Bolt	4.6	4.8	5.6	5.8	6.6	6.8	8.8	10.9	12.9	14.9
Recommended Grade of Nut	4	4	5	5	6	6	8	12	12	14

Notes

The single grade number for nuts indicates one tenth of the proof stress in kgf/mm^2 and corresponds with the bolt ultimate strength to which it is matched. It is permissible to use a higher strength grade nut than the matching bolt number. Grade 10.9 bolts are supplied with grade 12 nuts because grade 10 does not appear in the British Standard series.

Ordering example

Bolts M24 size 80 mm long, grade 8.8 with standard length of thread. With standard nut grade 8.8 and normal washer. All cadmium plated.

Bolts M24 × 80 to BS 3692 – 8.8 with standard nut and normal washer. All plated to BS 3382: Part 1.

Table 4.2   Dimensions of HSFG bolts.

Dimensions to BS 4395:Parts 1 & 2 (Nearest mm)

Nominal Size	Bolts and Nuts						Bolts			Nuts	Washers						Add to Grip for Length
	Pitch of Thread	Width Across Flats (max)	Width Across Girders (max)	Washer Face		Hex Head Depth (max)	Countersunk Head			Depth (max)	Round				Tapered		
				Dia. (max)	Depth		Dia	Flash	Min Ply		Diameter		Thickness (nom)	Clip	Overall Size	Inside Dia (nom)	
											Inside (nom)	Outside (max)					
d	–	s1	e	g	b	k	j	h	p	m	d1	d2	s	w	c	d3	–
(M12)	1.75	22	25	22	0.4	9	24	2	9	12	14	30	3	12	–	–	22
M16	2.0	27	31	27	0.4	11	32	2	9	16	18	37	3	14	38	18	26
M20	2.5	32	37	32	0.4	14	40	3	12	19	21	44	4	18	38	21	30
(M22)	2.5	36	42	36	0.4	15	44	3	13	20	23	50	4	19	45	23	34
M24	3.0	41	47	41	0.5	16	48	4	15	23	26	56	4	21	57	26	36
(M27)	3.0	46	53	46	0.5	18	54	4	16	25	29	60	4	23	57	29	39
M30	3.5	50	58	50	0.5	20	60	5	19	27	33	66	5	26	57	33	42
(M33)	3.5	55	64	55	0.5	22	66	5	20	30	36	75	–	29	57	36	45
M36	4.0	60	69	60	0.5	24	72	5	22	32	–	–	–	–	–	–	48

Mechanical Properties to BS 4395:Parts 1 & 2

Nominal Size	Tensile Stress Area	General Grade Part 1			Higher Grade Part 2		
		Proof Load	Yield Load	Ultimate Load	Proof Load	Yield Load	Ultimate Load
d	mm²	kN	kN	kN	kN	kN	kN
(M12)	84.3	49.4	53.3	69.6	–	–	–
M16	157	92.1	99.7	130	122.2	138.7	154.1
M20	245	144	155	203	190.4	216	240
(M22)	303	177	192	250	235.5	266	269.5
M24	358	207	225	292	274.6	312	345
(M27)	459	234	259	333	356	406	450
M30	561	286	313	406	435	495	550
(M33)	694	–	–	–	540	612	680
M36	817	418	445	591	–	–	–

Notes to Table 4.1

1. Commonly used sizes are underlined. Non-preferred sizes shown in brackets. Preferred larger diameters are M42, M56 and M64.

2. Bolt length (l) normally available in 5 mm increments up to 80 mm length and in 10 mm increments thereafter.

3. Sizes M16, M20, M24 and M27 up to 12.5 mm length may alternatively have a shorter thread length of 1½d, if so ordered. This may be required where the design does not allow the threaded portions across a shear plane.

4. BS 4190 covers black bolts of grades 4.6, 4.8 and 10.9. BS 3692 covers precision bolts in grades 4.6, 4.8, 5.6, 5.8, 6.6, 6.8, 8.8, 10.9, 12.9 and 14.9. Tolerances are closer and the maximum dimensions here quoted are slightly reduced.

Notes to Table 4.2

1. 'Add to Grip For Length' allows for nut, one flat round washer and sufficient thread protrusion beyond nut.

2. Bolt length (l) normally available in 5 mm increments up to 100 mm length and 10 mm increments thereafter. 10 mm increments are normally stocked by suppliers.

Table 4.3 Universal beams – to BS 4-1: 2005.

The dimension $N = [B - C + 6]$ to the nearest 2 mm above.　$n = \dfrac{D - d}{2}$ to the nearest 2 mm above.　$C = t/2 + 2$ mm to the nearest 1 mm.

Designation		Depth	Width	Thickness		Root Radius	Web	End Clearance	Notch		Hole spacings				Max. hole dia.	Area of Section	Surface Area
Serial size	Mass per metre	D	B	Web t	Flange T	r	Depth between fillets d	C	N	n	S₁	S₂	S₃	S₄		per metre
mm	kg	mm	mm	mm	mm	mm	mm	mm	mm	mm	mm	mm	mm	mm	mm	cm²	m²
914 × 419	388	920.5	420.5	21.5	36.6	24.1	799.0	13	208	62	140	140	75	290	24	494.5	3.404
	343	911.4	418.5	19.4	32.0	24.1	799.0	12	208	58						437.5	3.382
914 × 305	289	926.6	307.8	19.6	32.0	19.1	824.4	12	154	52	140	120	60	240	20	368.8	2.988
	253	918.5	305.5	17.3	27.9	19.1	824.4	11	154	48						322.8	2.967
	224	910.3	304.1	15.9	23.9	19.1	824.4	10	154	44						285.3	2.948
	201	903.0	303.4	15.2	20.2	19.1	824.4	10	154	40						256.4	2.932
838 × 292	226	850.9	293.8	16.1	26.8	17.8	761.7	10	148	46	140	–	–	–	24	288.7	2.791
	194	840.7	292.4	14.7	21.7	17.8	761.7	9	148	40						247.2	2.767
	176	834.9	291.6	14.0	18.8	17.8	761.7	9	148	38						224.1	2.754
762 × 267	197	769.6	268.0	15.6	25.4	16.5	685.8	10	136	42	140	–	–	–	24	250.8	2.530
	173	762.0	266.7	14.3	21.6	16.5	685.8	9	136	40						220.5	2.512
	147	753.9	265.3	12.9	17.5	16.5	685.8	8	136	36						188.1	2.493
686 × 254	170	692.9	255.8	14.5	23.7	15.2	615.0	9	130	40	140	–	–	–	24	216.6	2.333
	152	687.6	254.5	13.2	21.0	15.2	615.0	9	130	38						193.8	2.320
	140	683.5	253.7	12.4	19.0	15.2	615.0	8	130	36						178.6	2.310
	125	677.9	253.0	11.7	16.2	15.2	615.0	8	130	32						159.6	2.298
610 × 305	238	633.0	311.5	18.6	31.4	16.5	537.2	11	156	48	140	120	60	240	20	303.8	2.421
	179	617.5	307.0	14.1	23.6	16.5	537.2	9	156	42						227.9	2.381
	149	609.6	304.8	11.9	19.7	16.5	537.2	8	156	38						190.1	2.361
610 × 229	140	617.0	230.1	13.1	22.1	12.7	547.2	9	118	36	140	–	–	–	24	178.4	2.088
	125	611.9	229.0	11.9	19.6	12.7	547.2	8	118	34						159.6	2.075
	113	607.3	228.2	11.2	17.3	12.7	547.2	8	118	32						144.5	2.064
	101	602.2	227.6	10.6	14.8	12.7	547.2	7	118	28						129.2	2.053
533 × 210	122	544.6	211.9	12.8	21.3	12.7	476.5	8	108	34	140	–	–	–	24	155.8	1.872
	109	539.5	210.7	11.6	18.8	12.7	476.5	8	108	32						138.6	1.860
	101	536.7	210.1	10.9	17.4	12.7	476.5	7	108	30						129.3	1.853
	92	533.1	209.3	10.2	15.6	12.7	476.5	7	108	30						117.8	1.844
	82	528.3	208.7	9.6	13.2	12.7	476.5	7	108	26						104.4	1.833
457 × 191	98	467.4	192.8	11.4	19.6	10.2	407.9	8	100	30	90	–	–	–	24	125.3	1.650
	89	463.6	192.0	10.6	17.7	10.2	407.9	7	100	28						113.9	1.641
	82	460.2	191.3	9.9	16.0	10.2	407.9	7	100	28						104.5	1.633
	74	457.2	190.5	9.1	14.5	10.2	407.9	7	100	26						95.0	1.625
	67	453.6	189.9	8.5	12.7	10.2	407.9	6	100	24						85.4	1.617

Table 4.3 *Contd*

Designation		Depth	Width	Thickness		Root Radius	Web	End Clearance	Notch		Hole spacing	Max. hole dia.	Area of Section	Surface Area per metre
Serial size	Mass per metre	D	B	Web t	Flange T	r	Depth between fillets d	C	N	n	S_1			
mm	kg	mm	mm	mm	mm	mm	mm	mm	mm	mm	mm	mm	cm²	m²
457 × 152	82	465.1	153.5	10.7	18.9	10.2	406.9	7	80	30	90	20	104.5	1.493
	74	461.3	152.7	9.9	17.0	10.2	406.9	7	80	28			95.0	1.484
	67	457.2	151.9	9.1	15.0	10.2	406.9	7	80	26			85.4	1.474
	60	454.7	152.9	8.0	13.3	10.2	407.7	6	82	24			75.9	1.487
	52	449.8	152.4	7.6	10.9	10.2	407.7	6	82	22			66.5	1.476
406 × 178	74	412.8	179.7	9.7	16.0	10.2	360.5	7	94	28	90	24	95.0	1.493
	67	409.4	178.8	8.8	14.3	10.2	360.5	6	94	26			85.5	1.484
	60	406.4	177.8	7.8	12.8	10.2	360.5	6	94	24			76.0	1.476
	54	402.6	177.6	7.6	10.9	10.2	360.5	6	94	22			68.4	1.468
406 × 140	46	402.3	142.4	6.9	11.2	10.2	359.6	5	76	24	70	20	59.0	1.332
	39	397.3	141.8	6.3	8.6	10.2	359.6	5	76	20			49.4	1.320
356 × 171	67	364.0	173.2	9.1	15.7	10.2	312.2	7	92	26	90	24	85.4	1.371
	57	358.6	172.1	8.0	13.0	10.2	312.2	6	92	24			72.2	1.358
	51	355.6	171.5	7.3	11.5	10.2	312.2	6	92	22			64.6	1.351
	45	352.0	171.0	6.9	9.7	10.2	312.2	5	92	20			57.0	1.343
356 × 127	39	352.8	126.0	6.5	10.7	10.2	311.1	5	68	22	70	20	49.4	1.169
	33	348.5	125.4	5.9	8.5	10.2	311.1	5	68	20			41.8	1.160
305 × 165	54	310.9	166.8	7.7	13.7	8.9	265.6	6	88	24	90	24	68.4	1.245
	46	307.1	165.7	6.7	11.8	8.9	265.6	5	88	22			58.9	1.235
	40	303.8	165.1	6.1	10.2	8.9	265.6	5	88	20			51.5	1.227
305 × 127	48	310.4	125.2	8.9	14.0	8.9	264.6	6	68	24	70	20	60.8	1.079
	42	306.6	124.3	8.0	12.1	8.9	264.6	6	68	22			53.2	1.069
	37	303.8	123.5	7.2	10.7	8.9	264.6	6	68	20			47.5	1.062
305 × 102	33	312.7	102.4	6.6	10.8	7.6	275.8	5	56	20	54	12	41.8	1.006
	28	308.9	101.9	6.1	8.9	7.6	275.8	5	56	18			36.3	0.997
	25	304.8	101.6	5.8	6.8	7.6	275.8	5	56	16			31.4	0.988
254 × 146	43	259.6	147.3	7.3	12.7	7.6	218.9	6	80	22	70	20	55.1	1.069
	37	256.0	146.4	6.4	10.9	7.6	218.9	5	80	20			47.5	1.060
	31	251.5	146.1	6.1	8.6	7.6	218.9	5	80	18			40.0	1.050
254 × 102	28	260.4	102.1	6.4	10.0	7.6	225.0	5	56	18	54	12	36.2	0.900
	25	257.0	101.9	6.1	8.4	7.6	225.0	5	56	16			32.2	0.893
	22	254.0	101.6	5.8	6.8	7.6	225.0	5	56	16			28.4	0.887
203 × 133	30	206.8	133.8	6.3	9.6	7.6	172.3	5	72	18	70	20	38.0	0.912
	25	203.2	133.4	5.8	7.8	7.6	172.3	5	72	16			32.3	0.904

| Designation | | Depth | Width | Thickness | | Root Radius | Web | End Clearance | Notch | | Hole spacing | Max. hole dia. | Area of Section | Surface Area |
| Serial size | Mass per metre | D | B | Web t | Flange T | r | Depth between fillets d | C | N | n | S_1 | | | per metre |
mm	kg	mm	mm	mm	mm	mm	mm	mm	mm	mm	mm	mm	cm²	m²
203 × 102	23	203.2	101.6	5.2	9.3	7.6	169.4	5	56	18	54	12	29.0	0.789
178 × 102	19	177.8	101.6	4.7	7.9	7.6	146.8	4	58	16	54	12	24.2	0.740
152 × 89	16	152.4	88.9	4.6	7.7	7.6	121.8	4	52	16	50	–	20.5	0.638
127 × 76	13	127.0	76.2	4.2	7.6	7.6	96.6	4	46	16	40	–	16.8	0.537

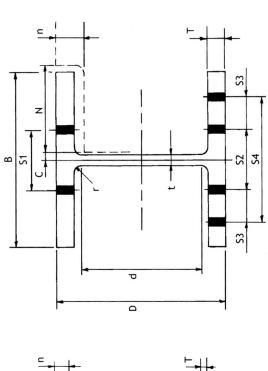

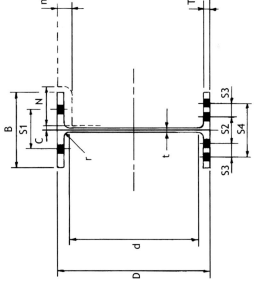

Table 4.4 Universal columns.

The Dimension $N = [B - C + 6]$ to the nearest 2 mm above.

$n = \dfrac{D - d}{2}$ to the nearest 2 mm above.

$C = t/2 + 2$ mm to the nearest 1 mm.

To BS 4-1: 2005.

Serial size (mm)	Mass per metre (kg)	Depth of Section D (mm)	Width of Section B (mm)	Web t (mm)	Flange T (mm)	Root Radius r (mm)	Web Depth between fillets d (mm)	End Clearance C (mm)	Notch N (mm)	Notch n (mm)	S₁ (mm)	S₂ (mm)	S₃ (mm)	S₄ (mm)	Max. hole dia. (mm)	Area of Section (cm²)	Surface Area per metre (m²)
356 × 406	634	474.7	424.1	47.6	77.0	15.2	290.1	26	198	94	140	140	75	290	24	808.1	2.525
	551	455.7	418.5	42.0	67.5	15.2	290.1	23	198	84						701.8	2.475
	467	436.6	412.4	35.9	58.0	15.2	290.1	20	198	74						595.5	2.425
	393	419.1	407.0	30.6	49.2	15.2	290.1	17	198	66						500.9	2.379
	340	406.4	403.0	26.5	42.9	15.2	290.1	15	198	60						432.7	2.346
	287	393.7	399.0	22.6	36.5	15.2	290.1	13	198	52						366.0	2.312
	235	381.0	395.0	18.5	30.2	15.2	290.1	11	198	46						299.8	2.279
356 × 368	202	374.7	374.4	16.8	27.0	15.2	290.1	10	188	44	140	140	75	290	24	257.9	2.187
	177	368.3	372.1	14.5	23.8	15.2	290.1	9	188	44						225.7	2.170
	153	362.0	370.2	12.6	20.7	15.2	290.1	8	188	36						195.2	2.154
	129	355.6	368.3	10.7	17.5	15.2	290.1	7	188	34						164.9	2.137
305 × 305	283	365.3	321.3	26.9	44.1	15.2	246.6	15	156	60	140	120	60	240	24	360.4	1.938
	240	352.6	317.9	23.0	37.7	15.2	246.6	13	156	54						305.6	1.905
	198	339.9	314.1	19.2	31.4	15.2	246.6	12	156	48						252.3	1.872
	158	327.2	310.6	15.7	25.0	15.2	246.6	10	156	42	140	120	60	240	20	201.2	1.839
	137	320.5	308.7	13.8	21.7	15.2	246.6	9	156	38						174.6	1.822
	118	314.5	306.8	11.9	18.7	15.2	246.6	8	156	34						149.8	1.806
	97	307.8	304.8	9.9	15.4	15.2	246.6	7	156	32						123.3	1.789
254 × 254	167	289.1	264.5	19.2	31.7	12.7	200.2	12	132	46	140	–	–	–	24	212.4	1.576
	132	276.4	261.0	15.6	25.3	12.7	200.2	10	132	40						168.9	1.543
	107	266.7	258.3	13.0	20.5	12.7	200.2	9	132	34						136.6	1.519
	89	260.4	255.9	10.5	17.3	12.7	200.2	7	132	32						114.0	1.502
	73	254.0	254.0	8.6	14.2	12.7	200.2	6	132	28						92.9	1.485
203 × 203	86	222.3	208.8	13.0	20.5	10.2	160.8	8	106	32	140	–	–	–	24	110.1	1.236
	71	215.9	206.2	10.3	17.3	10.2	160.8	7	106	28						91.1	1.218
	60	209.6	205.2	9.3	14.2	10.2	160.8	7	106	26						75.8	1.204
	52	206.2	203.9	8.0	12.5	10.2	160.8	6	106	24						66.4	1.194
	46	203.2	203.2	7.3	11.0	10.2	160.8	6	106	22						58.8	1.187
152 × 152	37	161.8	154.4	8.1	11.5	7.6	123.4	6	82	20	90	–	–	–	20	47.4	0.912
	30	157.5	152.9	6.6	9.4	7.6	123.4	5	82	18						38.2	0.900
	23	152.4	152.4	6.1	6.8	7.6	123.4	5	82	16						29.8	0.889

Table 4.5 Joists.

The Dimension $N = [B - C + 6]$ to the nearest 2 mm above.

$n = \dfrac{D - d}{2}$ to the nearest 2 mm above.

$C = [t/2 + 2]$ to the nearest 1 mm.

To BS 4-1: 2005.

Serial size mm	Mass per metre kg	Depth of Section D mm	Width of Section B mm	Web t mm	Flange T mm	Root radius r₁ mm	Toe radius r₂ mm	Inside Slope degrees	Depth d mm	End Clearance C mm	Notch N mm	Notch n mm	S mm	Max. hole dia. mm	Area of Section cm²	Surface Area per metre m²
254 × 203	81.85	254.0	203.2	10.2	19.9	19.6	9.7	8	166.6	7	106	45	140	24	104.4	1.193
254 × 114	37.20	254.0	114.3	7.6	12.8	12.4	6.1	8	199.1	6	62	30	65	16	47.4	0.882
203 × 152	52.09	203.2	152.4	8.9	16.5	15.5	7.6	8	133.4	7	80	40	90	20	66.4	0.911
152 × 127	37.20	152.4	127.0	10.4	13.2	13.5	6.6	8	94.5	7	68	35	70	20	47.5	0.722
127 × 114	29.76	127.0	114.3	10.2	11.5	12.4	4.8	8	71.9	7	62	30	65	16	37.3	0.620
127 × 114	26.79	127.0	114.3	7.4	11.4	9.9	5.0	8	79.2	6	62	25	65	16	34.1	0.635
127 × 76	16.37	127.0	76.2	5.6	9.6	9.4	4.6	8	86.4	5	44	25	40	–	21.0	0.498
114 × 114	26.79	114.3	114.3	9.5	10.7	14.2	3.2	8	61.1	7	62	30	65	16	34.4	0.600
102 × 102	23.07	101.6	101.6	9.5	10.3	11.1	3.2	8	54.0	7	54	25	54	12	29.4	0.528
89 × 89	19.35	88.9	88.9	9.5	9.9	11.1	3.2	8	45.2	7	48	25	50	–	24.9	0.460
76 × 76	12.65	76.2	76.2	5.1	8.4	9.4	4.6	8	38.1	5	40	20	40	–	16.3	0.403

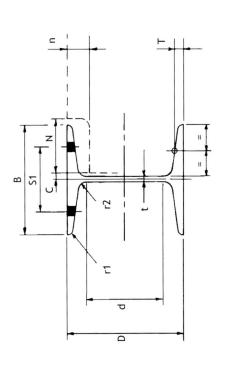

Table 4.6 Channels.

The Dimension $N = [B - C + 6]$ to the nearest 2 mm above. $n = \dfrac{D - d}{2}$ to the nearest 2 mm above. $C = [t + 2]$ to the nearest 1 mm.

To BS 4-1: 2005.

| Designation | | Depth | Width | Thickness | | Distance of y | Root radius | Toe radius | Inside Slope | Depth | | Notch | | | Max. hole dia. | Area of Section | Surface Area per metre |
| Serial size | Mass per metre | **D** | **B** | Web **t** | Flange **T** | **ey** | **r₁** | **r₂** | | **d** | **C** | **N** | **n** | **S** | | | |
mm	kg	mm	mm	mm	mm	cm	mm	mm	degrees	mm	mm	mm	mm	mm	mm	cm²	m²
432 × 102	65.54	431.8	101.6	12.2	16.8	2.32	15.2	4.8	5	362.5	14	94	36	55	20	83.49	1.21
381 × 102	55.10	381.0	101.6	10.4	16.3	2.52	15.2	4.8	5	312.6	12	96	36	55	20	70.19	1.11
305 × 102	46.18	304.8	101.6	10.2	14.8	2.66	15.2	4.8	5	239.3	12	96	34	55	20	58.83	0.96
305 × 89	41.69	304.8	88.9	10.2	13.7	2.18	13.7	3.2	5	245.4	12	84	30	50	20	53.11	0.92
254 × 89	35.74	254.0	88.9	9.1	13.6	2.42	13.7	3.2	5	194.7	11	84	30	50	20	45.52	0.82
254 × 76	28.29	254.0	76.2	8.1	10.9	1.86	12.2	3.2	5	203.9	10	74	26	45	20	36.03	0.774
229 × 89	32.76	228.6	88.9	8.6	13.3	2.53	13.7	3.2	5	169.9	11	84	30	50	20	41.73	0.770
229 × 76	26.06	228.6	76.2	7.6	11.2	2.00	12.2	3.2	5	177.8	10	74	26	45	20	33.20	0.725
203 × 89	29.78	203.2	88.9	8.1	12.9	2.65	13.7	3.2	5	145.2	10	86	30	50	20	37.94	0.720
203 × 76	23.82	203.2	76.2	7.1	11.2	2.13	12.2	3.2	5	152.4	9	74	26	45	20	30.34	0.675
178 × 89	26.81	177.8	88.9	7.6	12.3	2.76	13.7	3.2	5	121.0	10	86	30	50	20	34.15	0.671
178 × 76	20.84	177.8	76.2	6.6	10.3	2.20	12.2	3.2	5	128.8	9	74	26	45	20	26.54	0.625
152 × 89	23.84	152.4	88.9	7.1	11.6	2.86	13.7	3.2	5	96.9	9	86	28	50	20	30.36	0.621
152 × 76	17.88	152.4	76.2	6.4	9.0	2.21	12.2	2.4	5	105.9	8	76	24	45	20	22.77	0.575
127 × 64	14.90	127.0	63.5	6.4	9.2	1.94	10.7	2.4	5	84.0	8	62	22	35	16	18.98	0.476
102 × 51	10.42	101.6	50.8	6.1	7.6	1.51	9.1	2.4	5	65.8	8	50	18	30	10	13.28	0.379
76 × 38	6.72	76.2	38.1	5.1	6.8	1.19	7.6	2.4	5	45.8	7	38	16	22	10	8.56	0.282

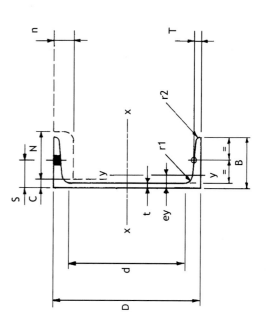

Table 4.7 Rolled steel angles: (a) equal (see p. 63).

Note: *Not included in BS EN 10056-1: 1999

To BS EN 10056-1: 1999

Designation Size B (mm)	Thickness t (mm)	Mass per metre (kg)	Area of section (cm²)	Distance of centre of gravity ex, ey (cm)	Recommended back marks S₁ (mm)	S₂ (mm)	S₃ (mm)	Max. dia. bolt (mm)
250 × 250	35	128.0	163.0	7.49	–	90	100	36
	32	118.0	150.0	7.38				
	28	104.0	133.0	7.23				
	25	93.6	119.0	7.12				
200 × 200	24	71.1	90.6	5.84	–	75	75	30
	20	59.9	76.3	5.68				
	18	54.2	69.1	5.60				
	16	48.5	61.8	5.52				
150 × 150	18	40.1	51.0	4.37	–	55	55	20
	15	33.8	43.0	4.25				
	12	27.3	34.8	4.12				
	10	23.0	29.3	4.03				
120 × 120	15	26.6	33.9	3.51	–	45	50	16
	12	21.6	27.5	3.40				
	10	18.2	23.2	3.31				
	8	14.7	18.7	3.23				
100 × 100	15	21.9	27.9	3.02	55	–	–	24
	12	17.8	22.7	2.90				
	10*	15.0	19.2	2.82				
	8	12.2	15.5	2.74				
90 × 90	12	15.9	20.3	2.66	50	–	–	24
	10	13.4	17.1	2.58				
	8	10.9	13.9	2.50				
	7	9.6	12.3	2.41				
	6	8.3	10.6	2.46				
80 × 80	10	11.9	15.1	2.34	45	–	–	20
	8	9.6	12.3	2.26				
	6	7.3	9.4	2.17				

Table 4.7 Rolled steel angles: (b) unequal (see p. 63).

To BS EN 10056-1: 1999

Designation		Mass per metre	Area of section	Distance of centre of gravity		Recommended back marks			Max. dia. bolt	
Size $D \times B$	Thickness t			e_x	e_y	S_1	S_2	S_3	For S_1	For $S_2 S_3$
mm	mm	kg	cm^2	cm	cm	mm	mm	mm	mm	mm
200 × 150	18	47.1	60.0	6.33	3.85					
	15	39.6	50.5	6.21	3.73	55	75	75	–	30
	12	32.0	40.8	6.08	3.61					
200 × 100	15	33.7	43.0	7.16	2.22					
	12	27.3	34.8	7.03	2.10	55	75	75	24	30
	10	23.0	29.2	6.93	2.01					
150 × 90	15	26.6	33.9	5.21	2.23					
	12	21.6	27.5	5.08	2.12	50	55	55	24	20
	10	18.2	23.2	5.00	2.04					
150 × 75	15	24.8	31.6	5.53	1.81					
	12	20.2	25.7	5.41	1.69	45	55	55	20	20
	10	17.0	21.6	5.32	1.61					
125 × 75	12	17.8	22.7	4.31	1.84					
	10	15.0	19.1	4.23	1.76	45	45	50	20	20
	8	12.2	15.5	4.14	1.68					
100 × 75	12	15.4	19.7	3.27	2.03					
	10	13.0	16.6	3.19	1.95	45	55	–	20	20
	8	10.6	13.5	3.10	1.87					
100 × 65	10	12.3	15.6	3.36	1.63					
	8	9.94	12.7	3.27	1.55	35	55	–	–	20
	7	8.77	11.2	3.23	1.51					
80 × 60	8	8.32	10.6	2.55	1.56					
	7	7.34	9.35	2.50	1.52	35	45	–	16	20
	6	6.34	8.08	2.46	1.48					
75 × 50	8	7.41	9.44	2.53	1.29	28	45	–	12	20
	6	5.67	7.22	2.44	1.21					
65 × 50	8	6.76	8.61	2.12	1.37					
	6	5.18	6.59	2.04	1.30	28	35	–	12	20
	5	4.36	5.55	2.00	1.26					
60 × 30	6	4.00	5.09	2.20	0.73	20	35	–	–	16
	5	3.38	4.30	2.16	0.68					
40 × 25	4	1.92	2.45	1.36	0.62	15	23	–	–	12

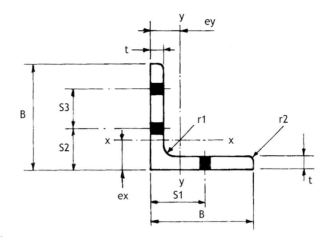

Equal angles (a)

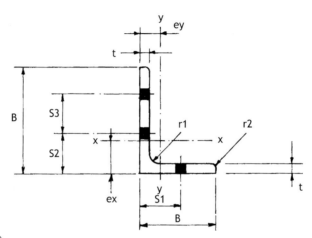

Unequal angles (b)

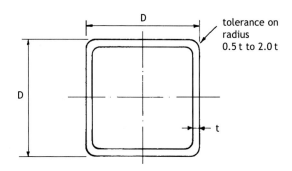

Table 4.8 Square hollow sections.

* Thickness not included in BS EN 10210-2: 2006

Designation		Mass per metre M	Area of section A	Surface area per metre	Designation		Mass per metre M	Area of section A	Surface area per metre
Size D x D	Thickness t				Size D x D	Thickness t			
mm	mm	kg	cm²	m²	mm	mm	kg	cm²	m²
20 x 20	2.0	1.12	1.42	0.076	120 x 120	5.0	18.0	22.9	0.469
	2.5*	1.35	1.72	0.075		6.3	22.3	28.5	0.466
25 x 25	2.0*	1.43	1.82	0.096		8.0	27.9	35.5	0.463
	2.5*	1.74	2.22	0.095		10.0	34.2	43.5	0.459
	3.2*	2.15	2.74	0.093	140 x 140	5.0	21.1	26.9	0.549
30 x 30	2.5*	2.14	2.72	0.115		6.3	26.3	33.5	0.546
	3.0*	2.51	3.20	0.114		8.0	32.9	41.9	0.543
	3.2	2.65	3.38	0.113		10.0	40.4	51.5	0.539
40 x 40	2.5*	2.92	3.72	0.155	150 x 150	5.0	22.7	28.9	0.589
	3.0*	3.45	4.40	0.154		6.3	28.3	36.0	0.586
	3.2	3.66	4.66	0.153		8.0	35.4	45.1	0.583
	4.0	4.46	5.68	0.151		10.0	43.6	55.5	0.579
50 x 50	2.5*	3.71	4.72	0.195		12.5	53.4	68.0	0.573
	3.0*	4.39	5.60	0.194		16.0	66.4	84.5	0.566
	3.2	4.66	5.94	0.193	180 x 180	6.3	34.2	43.6	0.706
	4.0	5.72	7.28	0.191		8.0	43.0	54.7	0.703
	5.0	6.97	8.88	0.189		10.0	53.0	67.5	0.699
60 x 60	3.0*	5.34	6.80	0.234		12.5	65.2	83.0	0.693
	3.2	5.67	7.22	0.233		16.0	81.4	104	0.686
	4.0	6.97	8.88	0.231	200 x 200	6.3	38.2	48.6	0.786
	5.0	8.54	10.9	0.229		8.0	48.0	61.1	0.783
70 x 70	3.0*	6.28	8.00	0.274		10.0	59.3	75.5	0.779
	3.6	7.46	9.50	0.272		12.5	73.0	93.0	0.773
	5.0	10.1	12.9	0.269		16.0	91.5	117	0.766
80 x 80	3.0*	7.22	9.20	0.314	250 x 250	6.3	48.1	61.2	0.986
	3.6	8.59	10.9	0.312		8.0	60.5	77.1	0.983
	5.0	11.7	14.9	0.309		10.0	75.0	95.5	0.979
	6.3	14.4	18.4	0.306		12.5	92.6	118	0.973
90 x 90	3.6	9.72	12.4	0.352		16.0	117	149	0.966
	5.0	13.3	16.9	0.349	300 x 300	10.0	90.7	116	1.18
	6.3	16.4	20.9	0.346		12.5	112	143	1.17
100 x 100	4.0	12.0	15.3	0.391		16.0	142	181	1.17
	5.0	14.8	18.9	0.389	350 x 350	10.0	106	136	1.38
	6.3	18.4	23.4	0.386		12.5	132	168	1.37
	8.0	22.9	29.1	0.383		16.0	167	213	1.37
	10.0	27.9	35.5	0.379	400 x 400	10.0	122	156	1.58
						12.5	152	193	1.57
						16.0	192	245	1.57

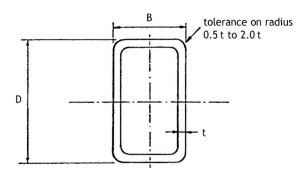

Table 4.9 Rectangular hollow sections.

* Thickness not included in BS EN 10210-2: 2006

Designation		Mass per metre	Area of section	Surface area per metre	Designation		Mass per metre	Area of section	Surface area per metre
Size D x B	Thickness t	M	A		Size D x B	Thickness t	M	A	
mm	mm	kg	cm²	m²	mm	mm	kg	cm²	m²
50 x 25	2.5*	2.72	3.47	0.145	**150 x 100**	5.0	18.7	23.9	0.489
	3.0*	3.22	4.10	0.144		6.3	23.3	29.7	0.486
	3.2*	3.41	4.34	0.143		8.0	29.1	37.1	0.483
50 x 30	2.5*	2.92	3.72	0.155		10.0	35.7	45.5	0.479
	3.0*	3.45	4.40	0.154	**160 x 80**	5.0	18.0	22.9	0.469
	3.2	3.66	4.66	0.153		6.3	22.3	28.5	0.466
60 x 40	2.5*	3.71	4.72	0.195		8.0	27.9	35.5	0.463
	3.0*	4.39	5.60	0.194		10.0	34.2	43.5	0.459
	3.2	4.66	5.94	0.193	**200 x 100**	5.0	22.7	28.9	0.589
	4.0	5.72	7.28	0.191		6.3	28.3	36.0	0.586
80 x 40	3.0*	5.34	6.80	0.234		8.0	35.4	45.1	0.583
	3.2	5.67	7.22	0.233		10.0	43.6	55.5	0.579
	4.0	6.97	8.88	0.231		12.5	53.4	68.0	0.573
90 x 50	3.0*	6.28	8.00	0.274		16.0	66.4	84.5	0.566
	3.6	7.46	9.50	0.272	**250 x 150**	6.3	38.2	48.6	0.785
	5.0	10.1	12.9	0.269		8.0	48.0	61.1	0.783
100 x 50	3.0*	6.75	8.60	0.294		10.0	59.3	75.5	0.779
	3.2	7.18	9.14	0.293		12.5	73.0	93.0	0.773
	4.0	8.86	11.3	0.291		16.0	91.5	117	0.766
	5.0	10.9	13.9	0.289	**300 x 200**	6.3	48.1	61.2	0.986
	6.3*	13.4	17.1	0.286		8.0	60.5	77.1	0.983
100 x 60	3.0*	7.22	9.20	0.314		10.0	75.0	95.5	0.979
	3.6	8.59	10.9	0.312		12.5	92.6	118	0.973
	5.0	11.7	14.9	0.309		16.0	117	149	0.966
	6.3	14.4	18.4	0.306	**400 x 200**	10.0	90.7	116	1.18
120 x 60	3.6	9.72	12.4	0.352		12.5	112	143	1.17
	5.0	13.3	16.9	0.349		16.0	142	181	1.17
	6.3	16.4	20.9	0.346	**450 x 250**	10.0	106	136	1.38
120 x 80	5.0	14.8	18.9	0.389		12.5	132	168	1.37
	6.3	18.4	23.4	0.386		16.0	167	213	1.37
	8.0	22.9	29.1	0.383					
	10.0	27.9	35.5	0.379					

Table 4.10 Circular hollow sections.

* Thickness not included in BS EN 10210-2: 2006

Designation		Mass per metre M	Area of section A	Surface area per metre
Outside diameter D	Thickness t			
mm	mm	kg	cm²	m²
21.3	3.2	1.43	1.82	0.067
26.9	3.2	1.87	2.38	0.085
33.7	2.6	1.99	2.54	0.106
	3.2	2.41	3.07	0.106
	4.0	2.93	3.73	0.106
42.4	2.6	2.55	3.25	0.133
	3.2	3.09	3.94	0.133
	4.0	3.79	4.83	0.133
48.3	3.2	3.56	4.53	0.152
	4.0	4.37	5.57	0.152
	5.0	5.34	6.80	0.152
60.3	3.2	4.51	5.74	0.189
	4.0	5.55	7.07	0.189
	5.0	6.82	8.69	0.189
76.1	3.2	5.75	7.33	0.239
	4.0	7.11	9.06	0.239
	5.0	8.77	11.2	0.239
88.9	3.2	6.76	8.62	0.279
	4.0	8.38	10.7	0.279
	5.0	10.3	13.2	0.279
114.3	3.6	9.83	12.5	0.359
	5.0	13.5	17.2	0.359
	6.3	16.8	21.4	0.359
139.7	5.0	16.6	21.2	0.439
	6.3	20.7	26.4	0.439
	8.0	26.0	33.1	0.439
	10.0	32.0	40.7	0.439
168.3	5.0	20.1	25.7	0.529
	6.3	25.2	32.1	0.529
	8.0	31.6	40.3	0.529
	10.0	39.0	49.7	0.529

Designation		Mass per metre M	Area of section A	Surface area per metre
Outside diameter D	Thickness t			
mm	mm	kg	cm²	m²
193.7	5.0*	23.3	29.6	0.609
	5.4	25.1	31.9	0.609
	6.3	29.1	37.1	0.609
	8.0	36.6	46.7	0.609
	10.0	45.3	57.7	0.609
	12.5	55.9	71.2	0.609
	16.0	70.1	89.3	0.609
219.1	5.0*	26.4	33.6	0.688
	6.3	33.1	42.1	0.688
	8.0	41.6	53.1	0:688
	10.0	51.6	65.7	0.688
	12.5	63.7	81.1	0.688
	16.0	80.1	102	0.688
	20.0	98.2	125	0.688
244.5	6.3	37.0	47.1	0.768
	8.0	46.7	59.4	0.768
	10.0	57.8	73.7	0.768
	12.5	71.5	91.1	0.768
	16.0	90.2	115	0.768
	20.0	111	141	0.768
273	6.3	41.4	52.8	0.858
	8.0	52.3	66.6	0.858
	10.0	64.9	82.6	0.858
	12.5	80.3	102	0.858
	16.0	101	129	0.858
	20.0	125	159	0.858
	25.0	153	195	0.858
323.9	6.3*	49.3	62.9	1.02
	8.0	62.3	79.4	1.02
	10.0	77.4	98.6	1.02
	12.5	96.0	122	1.02
	16.0	121	155	1.02
	20.0	150	191	1.02
	25.0	184	235	1.02

Table 4.10 *Contd*

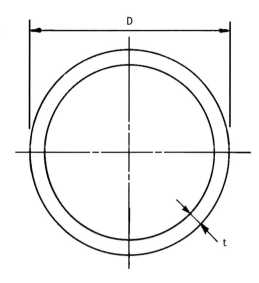

Designation		Mass per metre M	Area of section A	Surface area per metre
Outside diameter D	Thickness t			
mm	mm	kg	cm²	m²
355.6	8.0	68.6	87.4	1.12
	10.0	85.2	109	1.12
	12.5	106	135	1.12
	16.0	134	171	1.12
	20.0	166	211	1.12
	25.0	204	260	1.12
406.4	10.0	97.8	125	1.28
	12.5	121	155	1.28
	16.0	154	196	1.28
	20.0	191	243	1.28
	25.0	235	300	1.28
	32.0	295	376	1.28
457	10.0	110	140	1.44
	12.5	137	175	1.44
	16.0	174	222	1.44
	20.0	216	275	1.44
	25.0	266	339	1.44
	32.0	335	427	1.44
	40.0	411	524	1.44
508	10.0*	123	156	1.60
	12.5*	153	195	1.60
	16.0	194	247	1.60

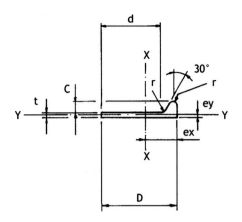

Table 4.11 Metric bulb flats.

Designation	Depth	Thickness	Bulb Height	Bulb Radius	Area of section	Mass per metre	Surface Area	Centroid	Moment of inertia	Modulus
	b	t	c	r1	A			ex	Ixx	Zxx
Size mm	mm	mm	mm	mm	cm²	kg/m	m²/m	cm	cm⁴	cm³
120 × 6	120	6	17	5	9.31	7.31	0.276	7.20	133	18.4
7		7	17	5	10.5	8.25	0.278	7.07	148	21.0
8		8	17	5	11.7	9.19	0.280	6.96	164	23.6
140 × 6.5	140	6.5	19	5.5	11.7	9.21	0.319	8.37	228	27.3
7		7	19	5.5	12.4	9.74	0.320	8.31	241	29.0
8		8	19	5.5	13.8	10.83	0.322	8.18	266	32.5
10		10	19	5.5	16.6	13.03	0.326	7.92	316	39.9
160 × 7	160	7	22	6	14.6	11.4	0.365	9.66	373	38.6
8		8	22	6	16.2	12.7	0.367	9.49	411	43.3
9		9	22	6	17.8	14.0	0.369	9.36	448	47.9
11.5		11.5	22	6	21.8	17.3	0.374	9.11	544	59.8
180 × 8	180	8	25	7	18.9	14.8	0.411	10.9	609	55.9
9		9	25	7	20.7	16.2	0.413	10.7	665	62.1
10		10	25	7	22.5	17.6	0.415	10.6	717	67.8
11.5		11.5	25	7	25.2	19.7	0.418	10.4	799	76.8
200 × 8.5	200	8.5	28	8	22.6	17.8	0.456	12.2	902	74.0
9		9	28	8	23.6	18.5	0.457	12.1	941	77.7
10		10	28	8	25.6	20.1	0.459	11.9	1020	85.0
11		11	28	8	27.6	21.7	0.461	11.8	1090	92.3
12		12	28	8	29.6	23.2	0.463	11.7	1160	99.6
220 × 9	220	9	31	9	26.8	21.0	0.501	13.6	1296	95.3
10		10	31	9	29.0	22.8	0.503	13.4	1400	105
11		11	31	9	31.2	24.5	0.505	13.2	1500	113
12		12	31	9	33.4	26.2	0.507	13.0	1590	122
240 × 9.5	240	9.5	34	10	31.2	24.4	0.546	14.8	1800	123
10		10	34	10	32.4	25.4	0.547	14.7	1860	126
11		11	34	10	34.9	27.4	0.549	14.6	2000	137
12		12	34	10	37.3	29.3	0.551	14.4	2130	148
260 × 10	260	10	37	11	36.1	28.3	0.593	16.2	2477	153
11		11	37	11	38.7	30.3	0.593	16.0	2610	162
12		12	37	11	41.3	32.4	0.595	15.8	2770	175
280 × 10.5	280	10.5	40	12	41.2	32.4	0.636	17.5	3223	184
11		11	40	12	42.6	33.5	0.637	17.4	3330	191
12		12	40	12	45.5	35.7	0.639	17.2	3550	206
13		13	40	12	48.4	37.9	0.641	17.0	3760	221
300 × 11	300	11	43	13	46.7	36.7	0.681	18.9	4190	222
12		12	43	13	49.7	39.0	0.683	18.7	4460	239
13		13	43	13	52.8	41.5	0.685	18.5	4720	256
320 × 11.5	320	11.5	46	14	52.6	41.2	0.727	20.2	5370	266
12		12	46	14	54.2	42.5	0.728	20.1	5530	274
13		13	46	14	57.4	45.0	0.730	19.9	5850	294
14		14	46	14	60.6	47.5	0.732	19.7	6170	313

Designation	Depth	Thickness	Bulb Height	Bulb Radius	Area of section	Mass per metre	Surface Area	Centroid	Moment of inertia	Modulus
	b	**t**	**c**	**r1**	**A**			**ex**	**Ixx**	**Zxx**
Size mm	mm	mm	mm	mm	cm^2	kg/m	m^2/m	cm	cm^4	cm^3
340 × 12	340	12	49	15	58.8	46.1	0.772	21.5	6760	313
13		13	49	15	62.2	48.8	0.774	21.3	7160	335
14		14	49	15	65.5	51.5	0.776	21.1	7540	357
15		15	49	15	69.0	54.2	0.778	20.9	7920	379
370 × 12.5	370	12.5	53.5	16.5	67.8	53.1	0.839	23.6	9213	390
13		13	53.5	16.5	69.6	54.6	0.840	23.5	9470	402
14		14	53.5	16.5	73.3	57.5	0.842	23.2	9980	428
15		15	53.5	16.5	77.0	60.5	0.844	23.0	10490	455
16		16	53.5	16.5	80.7	63.5	0.846	22.8	10980	481
400 × 13	400	13	58	18	77.4	60.8	0.907	25.8	12280	476
14		14	58	18	81.4	63.9	0.908	25.5	12930	507
15		15	58	18	85.4	67.0	0.910	25.2	13580	537
16		16	58	18	89.4	70.2	0.912	25.0	14220	568
430 × 14	430	14	62.5	19.5	89.7	70.6	0.975	27.7	16460	594
15		15	62.5	19.5	94.1	73.9	0.976	27.4	17260	628
17		17	62.5	19.5	103	80.6	0.980	26.9	18860	700
20		20	62.5	19.5	115	90.8	0.986	26.3	21180	804

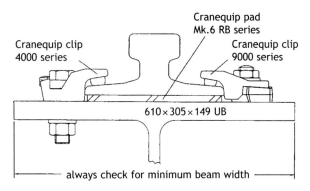

CROSS SECTION OF A55 RAIL ON RESILIENT PAD MOUNTING
CRANEQUIP WELDED AND BOLTED RAIL FASTENING

Table 4.12 Crane rails.

	F (mm)	K (mm)	H (mm)	Linear weight (kg/m)
A 45	125	45	55	22.1
A 55	150	55	65	31.8
A 65	175	65	75	43.1
A 75	200	75	85	56.2
A 100	200	100	95	74.3
A 120	220	120	105	100
A 150	220	150	150	150.3
28 BR	152	50	67	28.62
35 BR	160	58	76	35.38
56 CR	171	76	101.5	56.81
89 CR	178	102	114	89.81
CR 73	140	100	135	73.3
CR 100	155	120	150	100.2
MRS 87A	152.4	101.6	152.4	86.8
MRS 87B	152.4	102.4	152.4	86.8
MRS 125	180	120	180	125

Table 4.12 *Contd*

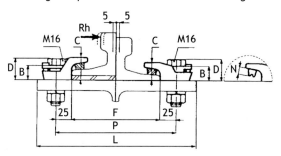

mounting with pad   mounting without pad

Series 4000

Reference	Dimensions (in mm)							Weight kg	Lat adjustment range (in mm)
	F	P	L (min)	B	C	D	N		
4116/10/27/10	175	225	281	22	27	32	10	0.34	10
4116/10/27/12	175	225	281	22	27	32	12	0.34	10
4116/10/34/10	175	225	281	22	34	32	10	0.40	10
4116/10/34/12	175	225	281	22	34	32	12	0.40	10
4120/15/28/10	175	245	320	21	28	34	10	0.58	15
4120/15/28/12	175	245	320	21	28	34	12	0.58	15
4120/15/35/10	175	245	320	21	35	34	10	0.60	15
4120/15/35/12	175	245	320	21	35	34	12	0.60	15
4120/15/40/08	175	245	320	21	40	34	08	0.64	15
4120/15/40/12	175	245	320	21	40	34	12	0.64	15
4124/15/35/10	175	245	320	24	35	40	10	0.70	15
4124/15/35/12	175	245	320	24	35	40	12	0.70	15
4124/15/40/08	175	245	320	24	40	40	08	0.80	15
4124/15/40/12	175	245	320	24	40	40	12	0.80	15

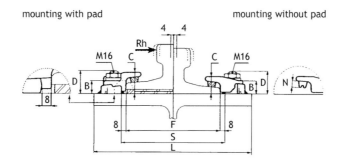

mounting with pad   mounting without pad

Series 9000

Reference	Dimensions							Weight kg	Lat adjustment range (in mm)
	F	S	L (min)	B	C	D	N		
9116/08/29/12	175	191	295	23	29	38	12	0.65	8
9116/08/29/15	175	191	295	23	29	38	15	0.65	8
9116/08/37/12	175	191	295	23	37	38	12	0.725	8
9116/08/37/15	175	191	295	23	37	38	15	0.725	8
9116/10/25/10	175	191	295	22	25	38	10	0.60	10
9116/10/32/10	175	191	295	22	32	38	10	0.625	10
9120/12/33/10	175	195	327	30	33	57	10	1.06	12
9120/12/33/13	175	195	327	30	33	57	13	1.06	12
9120/12/33/17	175	195	327	30	33	57	17	1.06	12
9120/12/40/10	175	195	327	30	40	57	10	1.15	12
9120/12/40/13	175	195	327	30	40	57	13	1.15	12
9120/12/40/17	175	195	327	30	40	57	17	1.15	12
9120/12/47/10	175	195	327	30	47	57	10	1.25	12
9120/12/47/13	175	195	327	30	47	57	13	1.25	12
9120/12/47/17	175	195	327	30	47	57	17	1.25	12
9216/08/33/12	175	191	275	29	33	42	12	1.55	8
9216/08/33/17	175	191	275	29	33	42	17	1.56	8
9216/08/40/12	175	191	275	29	40	42	12	1.575	8
9216/08/40/17	175	191	275	29	40	42	17	1.585	8
9216/08/43/12	175	191	275	32	43	45	12	1.64	8
9216/08/43/17	175	191	275	32	43	45	17	1.65	8
9220/18/45/12	175	203	340	35	45	48	12	2.70	18
9220/18/45/17	175	203	340	35	45	48	17	2.70	18
9220/18/52/12	175	203	340	35	52	48	12	2.90	18
9220/18/52/17	175	203	340	35	52	48	17	2.90	18

Crane rail clips available from CRANEQUIP Ltd, The Cape Industrial Estate, Cattell Road, Warwick, CV34 4JN

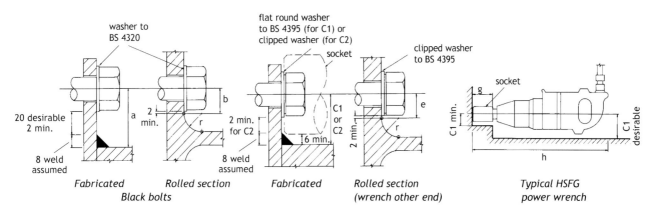

HSFG bolts

Table 4.13 Face clearances pitch and edge distance for bolts.

Diameter		Face clearance							Min. pitch		Minimum edge distance				HSFG wrench data		
Bolt	Hole	Black bolts			HSFG bolts				Black or HSFG	HSFG	BS 5950		BS 5400		Socket		Wrench
		a		b	c1		c2	e			Rolled Edge	Sheared	Black	HSFG	f	g	h
d	–															Length	Overall
Nom	Nom	Desirable	Minimum	Min	Desirable	Minimum	Min	Min	Minimum	Desirable	*	Sheared	*	HSFG	Dia		
(12)	14	40	22	14	56	35	22	14	30	45	18	20	17	21	57	65	420
16	18	43	25	17	56	35	24	16	40	50	23	26	22	27	57	65	420
20	22	47	29	21	56	35	28	20	50	55	28	31	27	33	57	65	420
(22)	24	48	30	22	56	39	29	21	55	60	30	34	29	36	65	65	420
24	26	50	32	24	56	45	31	23	60	65	33	37	32	39	78	90	411
(27)	30	53	35	27	64	45	39	25	68	70	38	42	36	45	78	115	550
30	33	56	38	30	64	55	36	28	75	80	42	47	40	50	97	115	550
(33)	36	58	40	32	64	55	39	31	83	85	45	51	44	54	97	115	550
36	39	61	43	35	64	55	42	34	90	90	49	55	47	59	97	115	550

Maximum pitch and edge distances for bolts		BS 5950	
	BS 5400	Corrosive	Non-corrosive
Edge distance	40 + 4t	40 + 4t	11t (M.S.) 9.7t (H.Y.S.)
Pitch — Along edge	100 + 4t or 200	16t or 200	–
Pitch — Any direction	32t or 300	16t or 200	–
Pitch — In direction of stress. Tension compression	16t or 200 / 12t or 200	14t	14t

Notes

1. See figure illustrating face clearances.
2. Dimensions of HSFG power wrench from *Structural fasteners and their application.*[6]
3. c1 = clearance required for HSFG wrench across.
   c2 = minimum clearance with clipped washer assuming wrench other end.
4. Minimum desirable pitch for HSFG bolts based on 2 mm clearance between socket and bolt head.
5. BS 5950 edge distances demand as:
   Rolled edge – rolled, machined flame cut, sawn or planed edge.
   Sheared   – sheared or hand plane cut edge and any edge.
* These values to be used with caution because reduction in bearing capacity occurs. Distance to value for HSFG bolts, t = minimum thickness of ply.

Table 4.14 Durbar floor plate.

Commercial quality – tensile range between 355 and 525 N/mm² with minimum range of 77 N/mm²
BS EN 10025: S275
Lloyds and other Shipbuilding Societies Specifications. Other specifications by arrangement.

Safe uniformly distributed load in kN/m² on plates simply supported on two sides. Extreme fibre stress: 165 N/mm².

Thickness on Plain mm.	Span in metres							
	0.6	0.8	1.0	1.2	1.4	1.6	1.8	2.0
4.5	12.29	6.97	4.47	3.10	2.28	1.77	1.37	1.12
6.0	22.06	12.41	7.94	5.52	4.04	3.12	2.44	1.98
8.0	39.24	22.12	14.09	9.83	7.18	5.54	4.34	3.56
10.0	61.22	34.45	22.00	15.33	11.22	8.67	6.78	5.55
12.5	95.82	53.91	34.44	23.99	17.56	13.57	10.61	8.70

The above safe loads include the weight of the plate. To avoid excessive deflection, stiffeners should be used for spans over 1.1 metres.

Safe uniformly distributed loads in kg/m² on plates simply supported on four sides. Extreme fibre stress: 16.537 kg/m².

Thickness on Plain mm.	Span in metres								Breadth Metres
	0.6	0.8	1.0	1.2	1.4	1.6	1.8	2.0	
4.5	24.86	16.36	14.05	13.21	12.85	12.70	12.58	12.53	0.60
		13.98	9.86	9.37	7.73	7.42	7.27	7.17	0.80
			8.95	6.63	5.65	5.16	4.90	4.76	1.00
6.0	44.13	29.02	24.93	23.44	22.82	22.48	22.33	22.24	0.60
		24.82	17.49	14.86	13.72	13.19	12.89	12.73	0.80
			15.89	11.77	10.02	9.15	8.70	8.44	1.00
				11.04	8.50	7.26	6.61	6.23	1.20
8.0	78.46	51.60	44.32	41.68	40.57	39.98	39.69	39.55	0.60
		44.13	31.10	26.41	24.40	23.44	22.93	22.62	0.80
			28.24	20.91	17.80	16.27	15.46	15.00	1.00
				19.60	15.11	12.89	11.74	11.08	1.20
					14.39	11.42	9.84	8.93	1.40
10.0	122.64	80.63	69.24	65.11	63.37	62.45	62.01	61.77	0.60
		68.95	48.59	41.28	38.13	36.62	35.83	35.36	0.80
			43.11	32.69	27.82	25.41	24.15	23.45	1.00
				30.63	23.61	20.15	18.34	17.28	1.20
					22.50	17.86	15.37	13.96	1.40
12.5	191.63	126.04	108.23	101.71	99.04	97.62	96.93	96.53	0.60
		107.73	75.93	64.50	59.60	57.22	56.01	55.26	0.80
			68.97	51.08	43.48	39.71	37.74	36.63	1.00
				47.88	36.92	31.51	28.67	27.05	1.20
					35.17	27.92	24.02	21.82	1.40

The above safe loads include the weight of the plate. The deflections on the larger spans should be checked and stiffeners used if found to be necessary.

Table 4.14 *Contd*

Standard sizes and weights

Width mm	Thickness Range on Plain mm				
1000					
1250	4.5	6.0	8.0	10.0	12.5
1500					
1750					
1830	—	6.0	8.0	10.0	12.5

Consideration will be given to requirements other than standard sizes where they represent a reasonable tonnage per size, i.e. in one length and one width. Lengths up to 10,000 mm can be supplied for plate 6 mm thick and over.

Weights per square metre or durbar plates

Thickness on Plain (mm)*	kg/m²
4.5	37.97
6.0	49.74
8.0	65.44
10.0	81.14
12.5	100.77

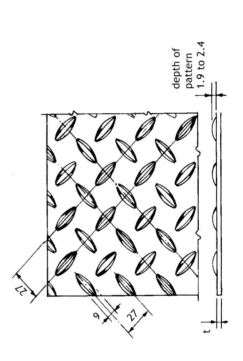

depth of pattern 1.9 to 2.4

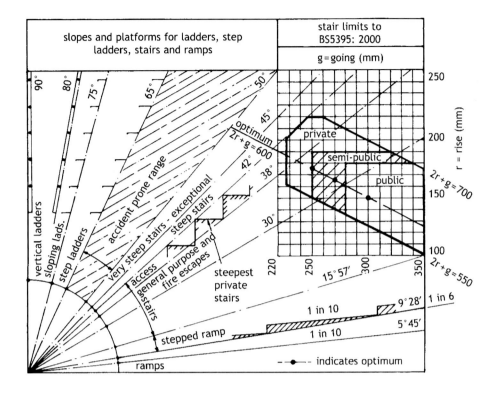

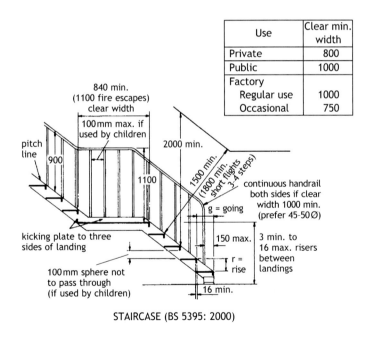

STAIRCASE (BS 5395: 2000)

Use	Clear min. width
Private	800
Public	1000
Factory	
Regular use	1000
Occasional	750

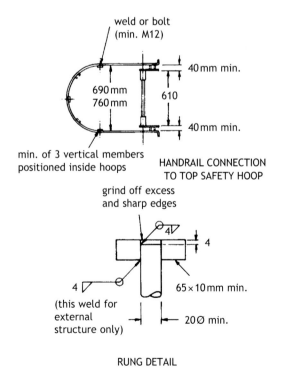

HANDRAIL CONNECTION
TO TOP SAFETY HOOP

RUNG DETAIL

Figure 4.1   Stairs, ladders and walkways.

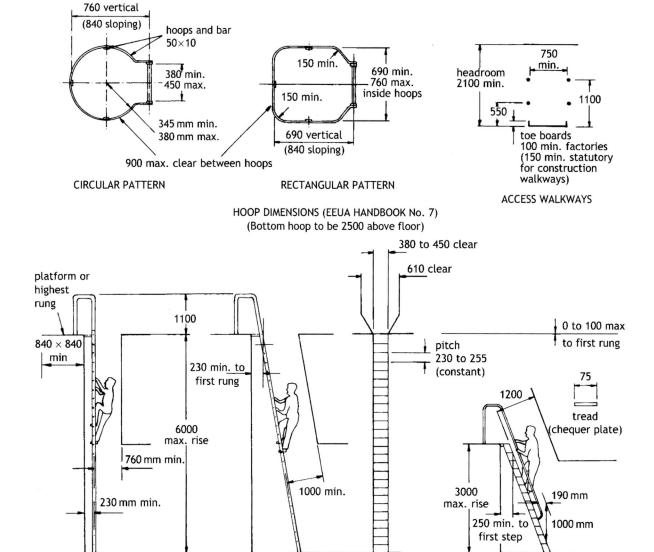

CIRCULAR PATTERN

RECTANGULAR PATTERN

ACCESS WALKWAYS

HOOP DIMENSIONS (EEUA HANDBOOK No. 7)
(Bottom hoop to be 2500 above floor)

VERTICAL LADDER

SLOPING LADDER
OVER 75°

STEP LADDER
65–75°

STAIRS, LADDERS AND STEP LADDERS (EEUA HANDBOOK No. 7)
Note: Ladders to be provided with hoops where rise exceeds 2–3 m

Figure 4.1 *Contd*

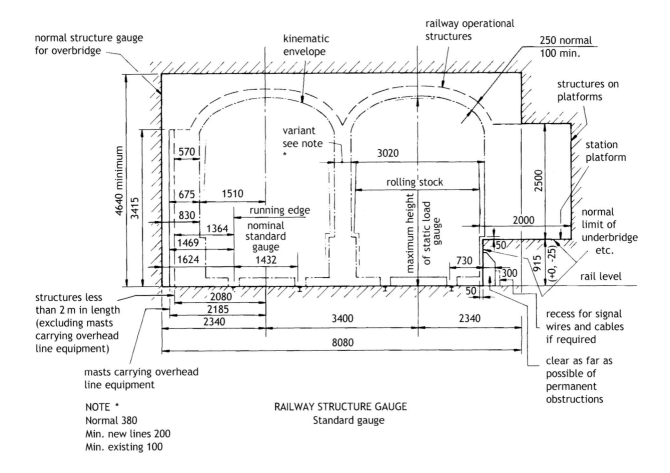

normal structure gauge for overbridge

kinematic envelope

railway operational structures

250 normal 100 min.

structures on platforms

station platform

4640 minimum

3415

570

variant see note *

3020

rolling stock

2500

normal limit of underbridge etc.

675   1510

830

running edge

nominal standard gauge

maximum height of static load gauge

2000

1469   1364

1624   1432

730

50

915 (+0, −25)

300

rail level

structures less than 2 m in length (excluding masts carrying overhead line equipment)

2080

2185

2340

3400

2340

50

recess for signal wires and cables if required

8080

masts carrying overhead line equipment

clear as far as possible of permanent obstructions

NOTE *
Normal 380
Min. new lines 200
Min. existing 100

RAILWAY STRUCTURE GAUGE
Standard gauge

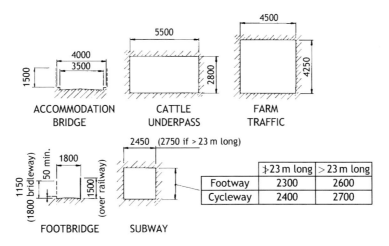

ACCOMMODATION BRIDGE

4000
3500
1500

CATTLE UNDERPASS

5500
2800

FARM TRAFFIC

4500
4250

FOOTBRIDGE

1150 (1800 bridleway)
50 min.
1800
1500 (over railway)

SUBWAY

2450 (2750 if > 23 m long)

	≯ 23 m long	> 23 m long
Footway	2300	2600
Cycleway	2400	2700

Figure 4.2 Highway and railway clearances.

ROADWAY HEADROOM H		
Type	New construction	Maintained
Overbridge	5300	5029
Footbridge	5300*	5029
Sign gantry	5700	5410

*5700 for >80 km/h

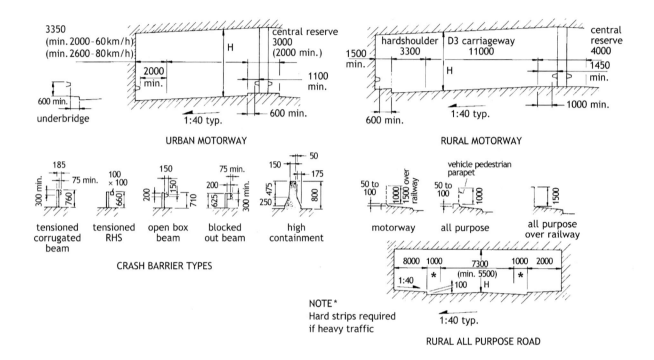

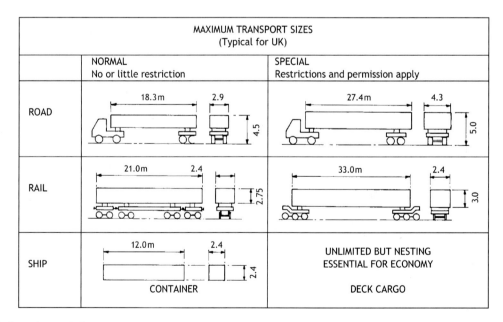

Figure 4.3   Maximum transport sizes.

WELDING SYMBOLS
These welding symbols are based upon BS 499 and are a selection of those most commonly used.
They should be used on engineer's & workshop drawings.

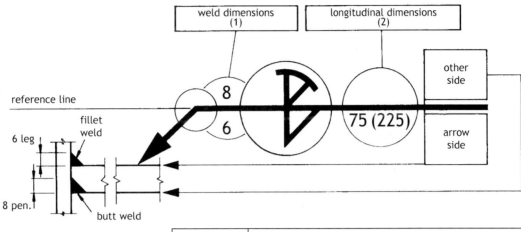

NOTES –
① Fillet – leg length
  Butt – penetration
    (no dimension indicates full
    penetration)
② Length of weld
  (no dimension indicates full length)
③ For other butt weld symbols see typical
  butt weld preparations
④ ←G→indicates ground flat and direction

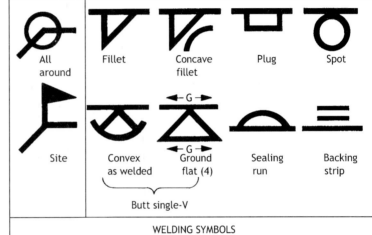

WELDING SYMBOLS

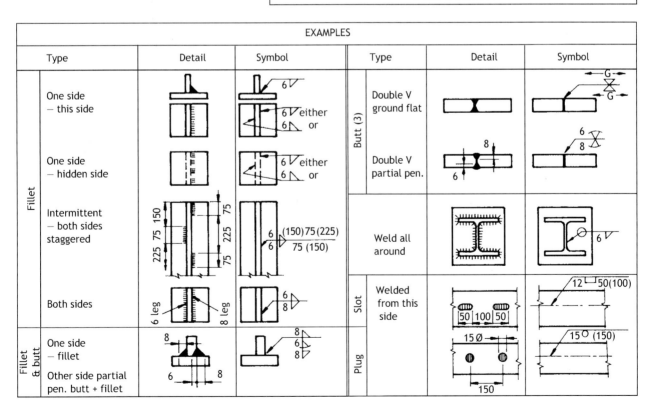

Figure 4.4 Weld symbols.

TYPICAL BUTT WELD PREPARATIONS — full penetration

These details are a typical selection only conforming with the recommended preparations in BS EN 1011-1: 2009 and BS EN 1011-2: 2001. Weld preparations should not be detailed on engineers drawings but are required on workshop drawings.

Weld & symbol	Detail	Thickness T	Gap G	Angle a	Root face R
		mm	mm		mm
Open square butt 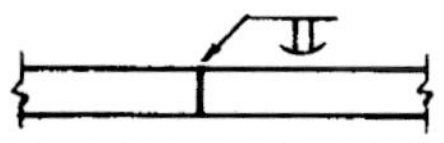	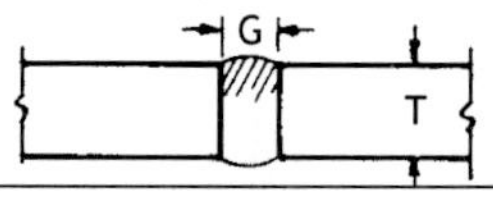	0–3 3–6	0–3 3	– –	– –
Open square butt backed 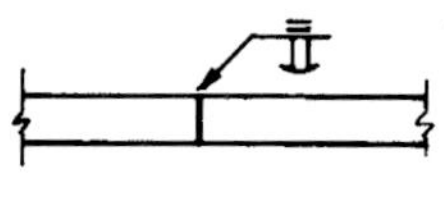	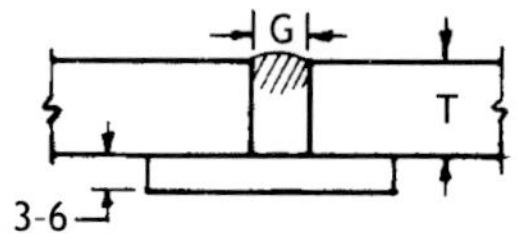	3–5 5–8 8–16	6 8 10	– – –	– – –
Single V butt 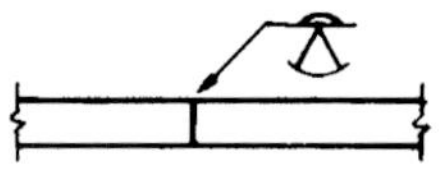	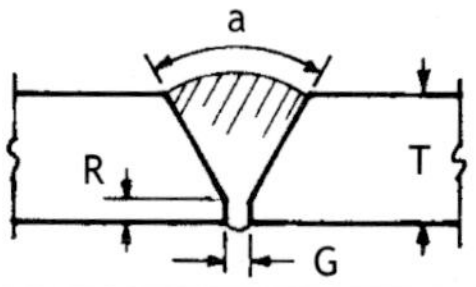	5–12 > 12	2 2	60° 60°	1 2
Single V butt backed 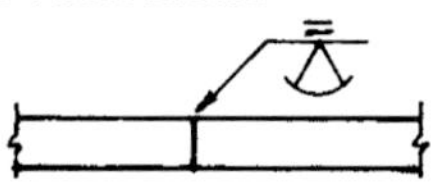	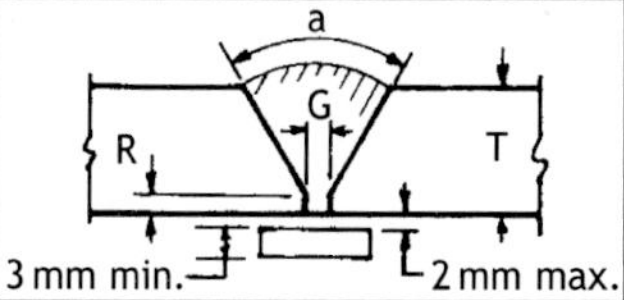	> 10	6 10	45° 20°	0 0
Double V butt 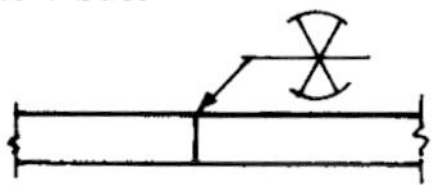	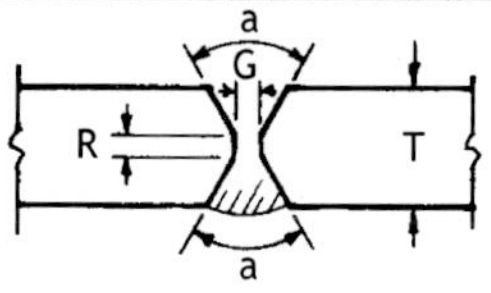	> 12	3	60°	2
Asymmetric double V butt 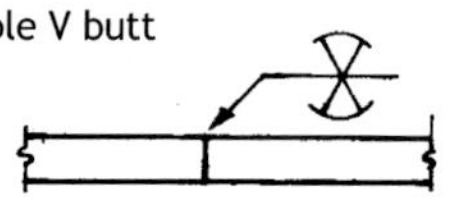	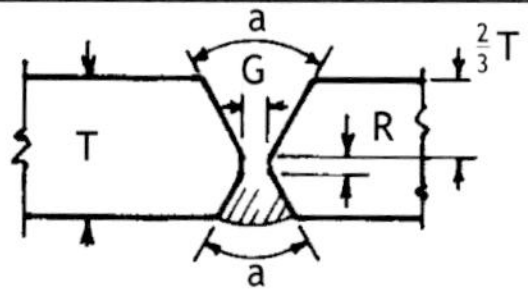	> 12	3	60°	2
Single J butt 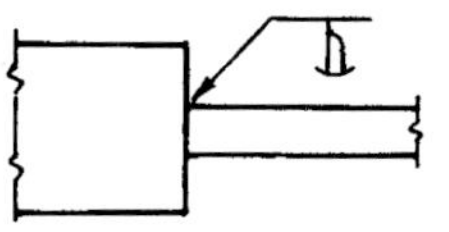	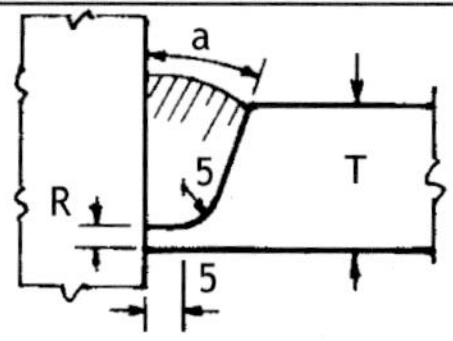	> 20	–	20°	5
Single U butt 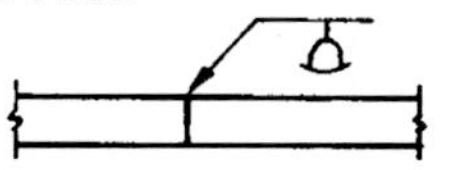	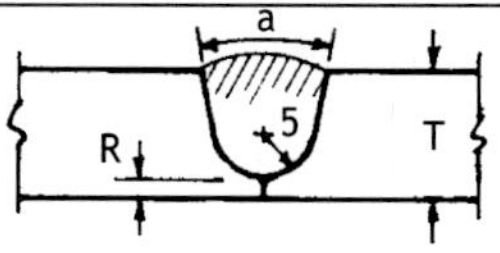	> 20	–	20°	5
Single bevel butt 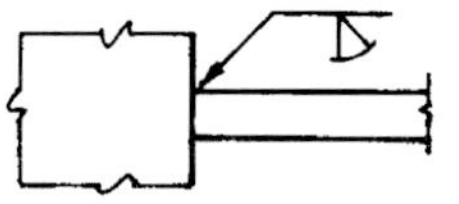	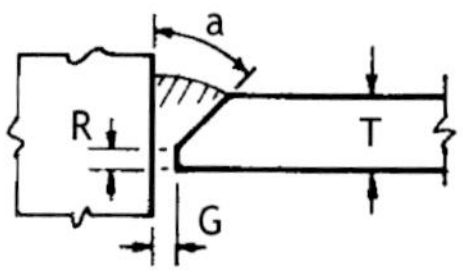	5–12 > 12	3 3	45° 45°	1 2
Double bevel butt 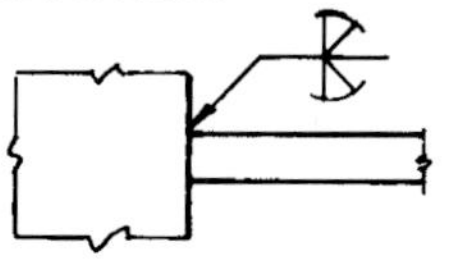	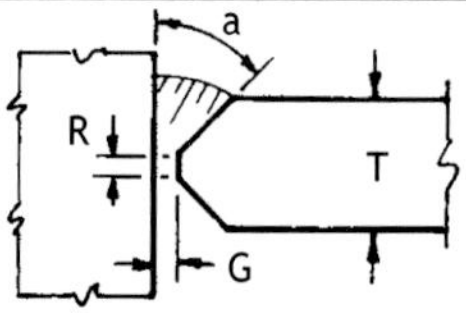	> 12	3	45°	2

Figure 4.5 Typical weld preparations.

TYPICAL PREPARATIONS FOR HOLLOW SECTIONS

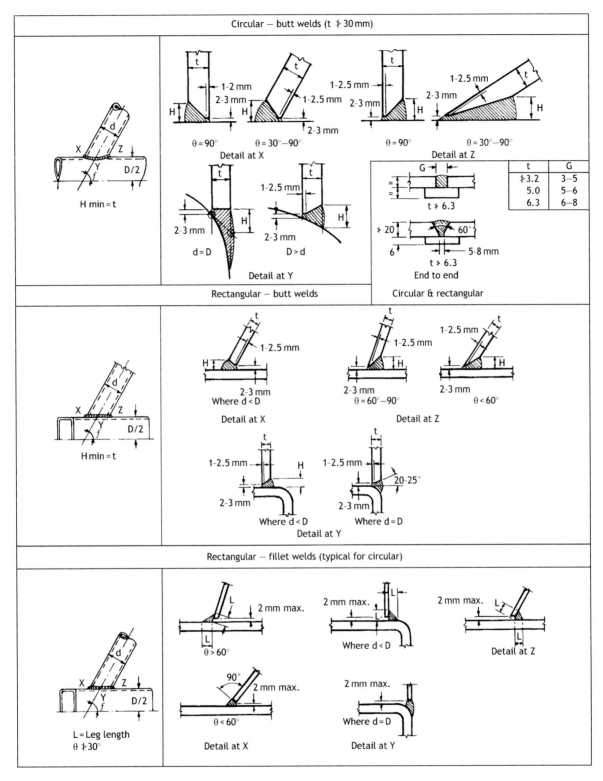

Figure 4.5 *Contd*

Table 4.15 Plates supplied in the 'Normalised' condition.
The information in this table must be read in conjunction with the explanatory notes at the bottom of the page.
Figures within the table are maximum lengths in metres.

Plate Gauge (mm)	≥1220 ≤1250	>1250 ≤1300	>1300 <1500	≥1500 ≤1600	>1600 ≤1750	≥1750 ≤1800	>1800 ≤2000	>2000 ≤2100	>2100 ≤2250	>2250 ≤2500	>2500 ≤2750	>2750 ≤3000	>3000 ≤3050	>3050 ≤3250	>3250 ≤3460	>3460 ≤3500	>3500 ≤3750
6	12.00	13.50	13.50	13.50	13.50	13.50	13.50	13.50	13.50	12.50	12.50	12.50	12.50	12.50	12.50		
7	12.00	13.50	13.50	13.50	13.50	13.50	13.50	13.50	13.50	13.50	13.50	13.50	13.50	13.50	13.50	13.50	
8	13.50	13.50	13.50	13.50	13.50	13.50	13.50	13.50	13.50	13.50	13.50	13.50	13.50	13.50	13.50		
9	15.50	15.50	15.50	15.50	15.50	15.50	15.50	15.50	15.50	15.50	15.50	15.50	15.50	15.50	15.50		
10	15.50	15.50	15.50	15.50	15.50	15.50	15.50	15.50	15.50	15.50	15.50	15.50	15.50	15.50	15.50		
12	15.50	15.50	15.50	15.50	15.50	15.50	15.50	15.50	15.50	15.50	15.50	15.50	15.50	15.50	15.50		
12.5	15.50	15.50	17.00	17.00	17.00	17.00	17.00	17.00	17.00	17.00	17.00	17.00	17.00	17.00	15.50	15.50	15.00
15	15.50	15.50	17.00	17.00	17.00	17.00	17.00	17.00	17.00	17.00	17.00	17.00	17.00	17.00	15.50	15.50	15.00
20	15.50	15.50	17.00	17.00	17.00	17.00	17.00	17.00	17.00	17.00	17.00	17.00	17.00	17.00	15.50	15.50	15.00
25	15.50	15.50	17.00	17.00	17.00	17.00	17.00	17.00	17.00	17.00	17.00	17.00	17.00	17.00	17.00	17.00	15.00
30	15.50	15.50	17.00	17.00	17.00	17.00	17.00	17.00	17.00	17.00	17.00	17.00	17.00	17.00	17.00	17.00	15.00
35	15.50	15.50	17.00	17.00	17.00	17.00	17.00	17.00	17.00	17.00	17.00	17.80	17.50	16.40	15.40	15.20	14.20
40	15.50	15.50	17.00	17.00	17.00	17.00	17.00	17.00	17.00	17.00	17.00	15.60	15.30	14.40	13.50	13.30	12.40
45	14.70	14.10	17.00	17.00	17.00	17.00	17.00	17.00	17.00	16.60	15.10	13.80	13.60	12.80	12.00	11.80	11.10
50	13.20	12.70	17.00	17.00	16.70	17.00	17.00	17.00	16.60	14.90	13.60	12.40	12.20	11.50	10.80	10.70	9.90
55	12.00	11.50	17.00	16.60	15.20	17.00	17.00	16.20	15.10	13.60	12.30	11.30	11.10	10.40	9.80	9.70	9.00
60	11.00	10.60	17.00	15.20	13.90	17.00	15.60	14.80	13.80	12.40	11.30	10.40	10.20	9.60	9.00	8.90	8.30
65	10.10	9.70	17.00	14.00	12.80	16.00	14.40	13.70	12.80	11.50	10.40	9.60	9.40	8.80	8.30	8.20	7.60
70	9.40	9.00	17.00	13.00	11.90	14.80	13.30	12.70	11.80	10.70	9.70	8.90	8.70	8.20	7.70	7.60	7.10
75	8.80	8.40	17.00	12.20	11.10	13.80	12.40	11.80	11.00	9.90	9.00	8.30	8.10	7.60	7.20	7.10	6.60
80	8.20	7.90	17.00	11.40	10.40	13.00	11.70	11.10	10.40	9.30	8.50	7.80	7.60	7.20	6.70	6.60	6.20
90	8.20	8.00	15.60	10.10	9.20	11.50	10.40	9.90	9.20	8.30	7.50	6.90	6.80	6.40	6.00	5.90	5.50
100	7.40	7.20	14.00	9.10	8.30	10.40	9.30	8.90	8.30	7.40	6.80	6.20	6.10	5.70	5.40	5.30	4.90
110	6.70	6.50	12.50	8.30	7.60	9.40	8.50	8.10	7.50	6.80	6.10	5.60	5.50	5.20	4.90	4.80	4.50
120	6.10	6.00	11.70	7.60	6.90	8.60	7.80	7.40	6.90	6.20	5.60	5.20	5.10	4.80	4.50	4.40	
125	5.90	5.70	11.20	7.30	6.60	8.30	7.40	7.10	6.60	5.90	5.40	4.90	4.90	4.60	4.30	4.20	
130	5.60	5.50	10.80	7.00	6.40	8.00	7.20	6.80	6.40	5.70	5.20	4.80	4.70	4.40			
140	5.20	5.10	10.00	6.50	5.90	7.40	6.60	6.30	5.90	5.30	4.80	4.40	4.30				
150	4.90	4.80	9.30	6.10	5.50	6.90	6.20	5.90	5.50	4.90	4.50						

There will be occasions when sizes and grades may be manufactured which are not shown in the table. As a guide to the available plate sizes for an intermediate gauge (e.g. 18mm), please use the nearest gauge shown in the table (i.e. 20mm). For any specific requirement, please contact us.

▨ Minimum 1.6 tonnes plate weight based on single plates or multiples.

▨ **Longer lengths**
Plates >17.0 and ≤19.0 metres may be supplied by agreement at the time of initial enquiry.

Typical qualities

Structural steels	Boiler steels	Ship plate (see Note 7)
EN10025-2:2004 S275, S355	EN10028:2003 P295GH, P355	Normal strength structural grades eg.
ASTM A572 50	BS 1501:1980 151/161 224, 225	Lloyds A, NVA, GL-A.
EN10025-3:2004 S275, S355, S420, S460	ASTM 516 Gr 60/65/70	Structural high strength grades eg. DH/EH 36

Notes

1. Plates >100mm thick and/or >14.5 tonnes, please contact us for confirmation of availability.
2. Plates >80mm thick and <1500mm wide are available by arrangement, if ordered in even numbers.
3. Plates >12.5mm to ≤80mm and <1500mm wide may be available in longer lengths than those shown, by arrangement and if ordered in even numbers.
4. Plates >3750mm wide are available by arrangement only.
5. Single plates ≤2000mm wide, please contact us for confir mation of availability.
6. Plates for structural, boiler & pressure vessel applications are also available to the requirements of European, ISO, ASTM, ASME, other national and customer standards.
7. Plates for ship construction are available to the requirements of major Classification Societies such as Lloyds Register, American Bureau of Shipping, Det Norske Veritas, Korean Register of Shipping, Germanischer Lloyd, RINA and Bureau Veritas.

Table 4.16 Plates supplied in the 'Normalised Rolled' and 'Thermo-Mechanically Controlled Rolled' condition.
The information in this table must be read in conjunction with the explanatory notes at the bottom of the page.
Figures within the table are maximum lengths in metres.

Plate Gauge (mm)	≥1220 ≤1250	>1250 ≤1300	>1300 1500	≥1500 ≤1600	>1600 ≤1750	>1750 ≤1800	>1800 ≤2000	>2000 ≤2100	>2100 ≤2250	>2250 ≤2500	>2500 ≤2750	>2750 ≤3000	>3000 ≤3050	>3050 ≤3250	>3250 ≤3460	>3460 ≤3500	>3500 ≤3750
								Plate Width (mm)									
8	13.50	13.50	13.50	13.50	13.50	13.50	13.50	13.50	13.50	13.50	13.50	13.50	13.50	13.50	13.50		
9	18.30	18.30	18.30	18.30	18.30	18.30	18.30	18.30	18.30	18.30	18.30	18.30	18.30	18.30	13.50		
10	18.30	18.30	18.30	18.30	18.30	18.30	18.30	18.30	18.30	18.30	18.30	18.30	18.30	18.30	13.50		
12	18.30	18.30	18.30	18.30	18.30	18.30	18.30	18.30	18.30	18.30	18.30	18.30	18.30	18.30	13.50		
12.5	18.30	18.30	18.30	18.30	18.30	18.30	18.30	18.30	18.30	18.30	18.30	18.30	18.30	18.30	18.30	15.00	
15	18.30	18.30	18.30	18.30	18.30	18.30	18.30	18.30	18.30	18.30	18.30	18.30	18.30	18.30	18.30	15.00	
20	18.30	18.30	18.30	18.30	18.30	18.30	18.30	18.30	18.30	18.30	18.30	18.30	18.30	18.30	18.30	15.00	
25	18.30	18.30	18.30	18.30	18.30	18.30	18.30	18.30	18.30	18.30	18.30	18.30	18.30	18.30	18.30	18.30	15.00
30	18.30	18.30	18.30	18.30	18.30	18.30	18.30	18.30	18.30	18.30	18.30	18.30	18.30	18.30	18.00	17.80	15.00
35	18.30	18.30	18.30	18.30	18.30	18.30	18.30	18.30	18.30	18.30	18.30	17.80	17.50	16.40	15.40	15.20	14.20
40	16.50	15.90	18.30	18.30	18.30	18.30	18.30	18.30	18.30	18.30	18.30	17.00	15.60	15.30	14.40	13.50	12.40
45	16.60	16.00	18.30	18.30	18.30	18.30	18.30	18.30	18.30	16.60	15.10	13.80	13.60	12.80	12.00	11.80	11.10
50	14.90	14.40	18.30	18.30	16.70	18.30	18.30	18.30	16.60	14.90	13.60	12.40	12.20	11.50	10.80	10.70	9.90
55	13.60	13.10	18.30	16.60	15.20	18.30	17.00	16.20	15.10	13.60	12.30	11.30	11.10	10.40	9.80	9.70	9.00
60	12.40	12.00	17.40	15.20	13.90	17.30	15.60	14.80	13.80	12.40	11.30	10.40	10.20	9.60	9.00	8.90	8.30

There will be occasions when sizes and grades may be manufactured which are not shown in the table. As a guide to the available plate sizes for an intermediate gauge (e.g. 18mm), please use the nearest gauge shown in the table (i.e. 20mm). For any specific requirement, please contact us.

The matrix can be used as a guide for plates supplied in the 'Thermo-Mechanically Controlled Rolled' (TMCR) condition. Note that all TMCR plates are based on a restricted size range, and confirmation of specific requirements need to be submitted as an enquiry to the technical department prior to order.

Minimum 1.6 tonnes plate weight based on single plates or multiples.

Longer lengths
Longer lengths (>18.3 metres) available by arrangement. Plates ≤27.4 metres may be supplied, depending on quality, gauge and width. Plates >27.4 to ≤31.5 metres may be available, for a limited quality and size range, by agreement at the time of initial enquiry.

Typical qualities

Structural steels
EN10025-2:2004 S355
EN10025-3:2004 (NR) S355, S420, S460
EN10025-4:2004 (TMCR) S355, S420, S460

Ship plate (see Note 7)
Higher strength structural grades, eg. DH36

Actis 360
Please refer to product brochure for details.

Weathering grades
EN10025-5:2004 S355 Including low CEV grades for improved weldability.

Notes
1. Plates >14.5 tonnes, please contact us for confirmation of availability.
2. Plates can also be supplied in the 'Thermo-Mechanically Controlled Rolled' condition by agreement.
3. Plates >40mm to ≤60mm and <1500mm wide are available by arrangement, if ordered in even numbers.
4. Plates >3750mm wide are available by arrangement only.
5. Single plates ≤2000mm wide, please contact us for confirmation of availability.
6. Plates for structural, boiler & pressure vessel applications are also available to the requirements of European, ISO, ASTM, ASME, other national and customer standards.
7. Plates for ship construction are available to the requirements of major Classification Societies such as Lloyds Register, American Bureau of Shipping, Det Norske Veritas, Korean Register of Shipping, Germanischer Lloyd, RINA and Bureau Veritas.

5 Connection Details

Following are sketch examples of typical connection details. These show the principles of some of the types of connection commonly used. Both simple and continuous connections are shown as applicable to beam/column structures. A typical workshop drawing of a roof lattice girder is included in figures 5.8 and 5.9. Sketches of steel/timber and steel/precast concrete connections are shown in figures 5.10 and 5.11 respectively.

Reference should also be made to a series of publications (see Further Reading, Design, (10), (11) and (12)) produced by BCSA and SCI which advocate the adoption of a range of connections to provide cost-effective design solutions. These books provide details of standardised simple and continuous connections, including capacity tables, dimensions for detailing and information on fasteners.

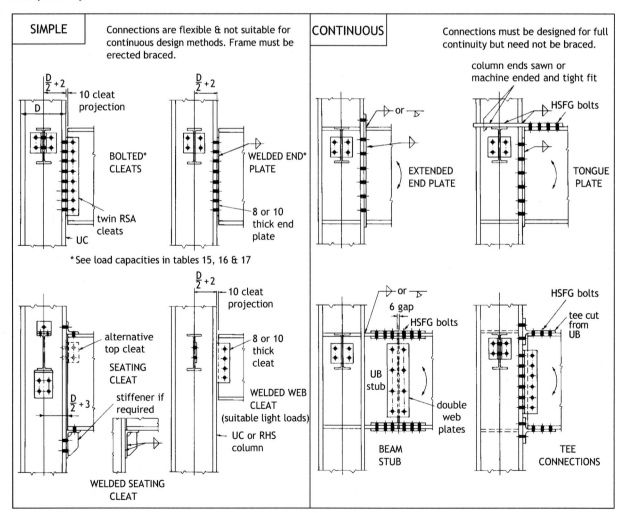

Figure 5.1 Typical beam/column connections.

Steel Detailers' Manual, Third Edition. Alan Hayward and Frank Weare. Third edition revised by Anthony Oakhill.
© 2011 Alan Hayward, Frank Weare and Anthony Oakhill. Published 2011 by Blackwell Publishing Ltd.

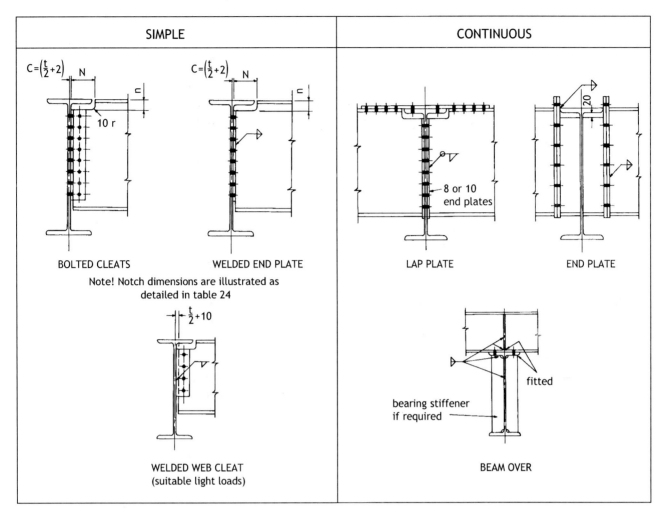

Figure 5.2 Typical beam/beam connections.

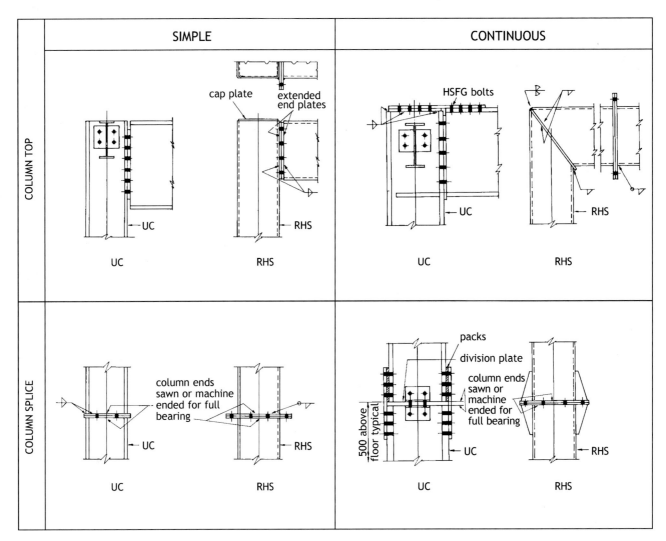

Figure 5.3 Typical column top and splice detail.

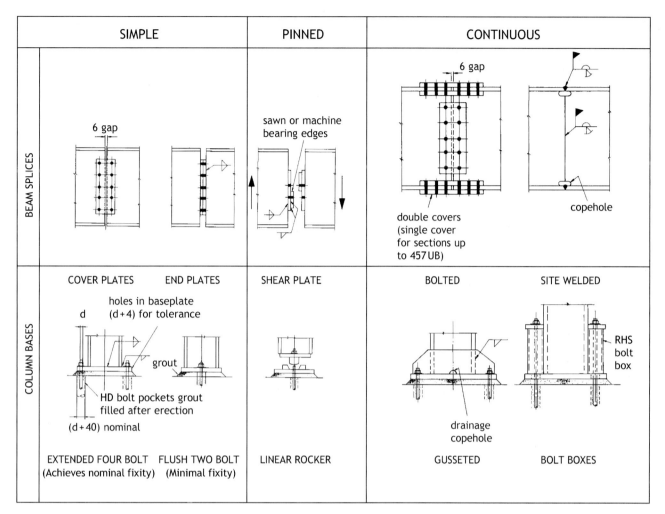

Figure 5.4 Typical beam splices and column bases.

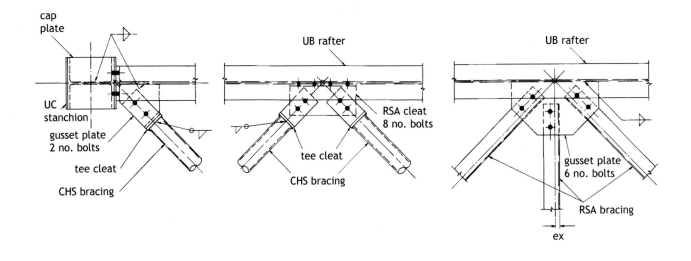

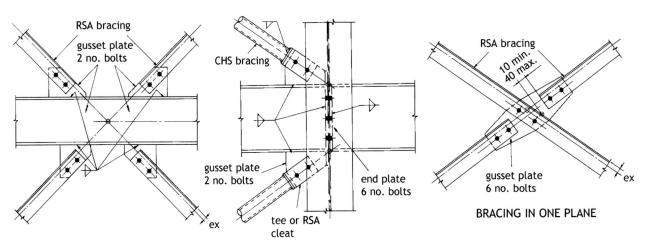

Figure 5.5 Typical bracing details.

RHS & CHS JOINTS

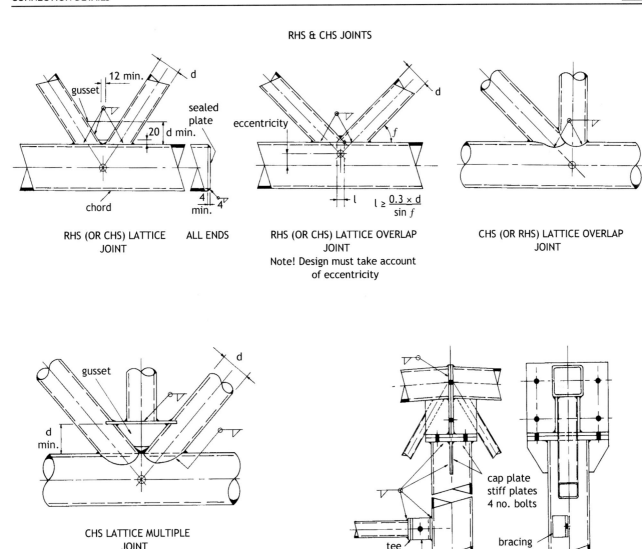

RHS (OR CHS) LATTICE
JOINT

ALL ENDS

RHS (OR CHS) LATTICE OVERLAP
JOINT
Note! Design must take account
of eccentricity

CHS (OR RHS) LATTICE OVERLAP
JOINT

CHS LATTICE MULTIPLE
JOINT

LATTICE GIRDER TO RHS COLUMN CAP

FLANGED SPLICE JOINT

NOTE: All hollow sections
to be fully sealed by welding

Figure 5.6 Typical hollow section connections.

WELDED

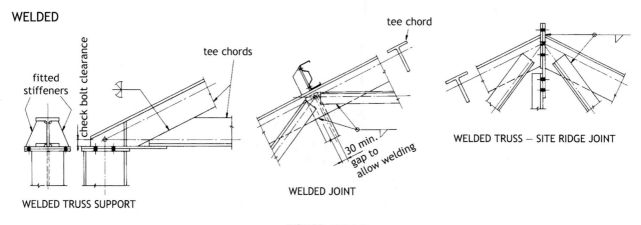

WELDED TRUSS SUPPORT

WELDED JOINT

WELDED TRUSS — SITE RIDGE JOINT

TRUSS JOINTS

BOLTED

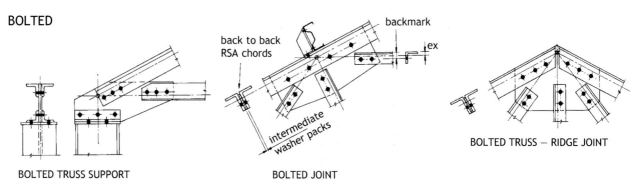

BOLTED TRUSS SUPPORT

BOLTED JOINT

BOLTED TRUSS — RIDGE JOINT

Figure 5.7   Typical truss details.

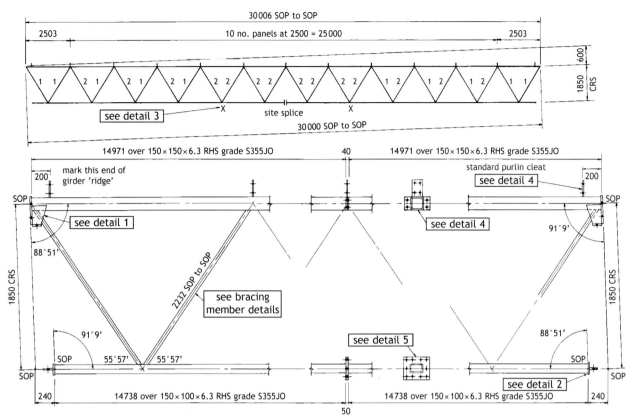

GENERAL NOTES:
1. All materials to EN 10025 grade S275JO UON.
2. All welds 6 fillet both sides of all joints UON.
   All hollow sections to be sealed by welding.
3. All bolts M20 (4.6).
4. All holes 22 dia.
5. Treatment — see spec.

20 GIRDERS REQD AS DRAWN MK T1

Figure 5.8 Workshop drawing of lattice girder – 1.

Shoe% 1 No. 350 × 12 PL × 310 long
      1 No. 100 × 12 PL × 350 long
      2 No. 150 × 10 PL (shaped)

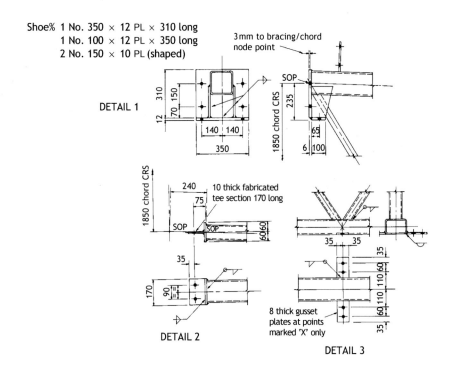

DETAIL 1

DETAIL 2

DETAIL 3

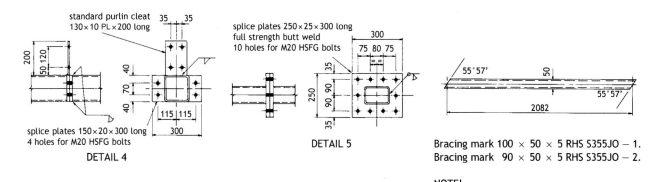

standard purlin cleat
130 × 10 PL × 200 long

splice plates 150 × 20 × 300 long
4 holes for M20 HSFG bolts

DETAIL 4

splice plates 250 × 25 × 300 long
full strength butt weld
10 holes for M20 HSFG bolts

DETAIL 5

Bracing mark 100 × 50 × 5 RHS S355JO − 1.
Bracing mark 90 × 50 × 5 RHS S355JO − 2.

NOTE!
For fabrication works with templating
facility this detail not necessary.

Figure 5.9 Workshop drawing of lattice girder − 2.

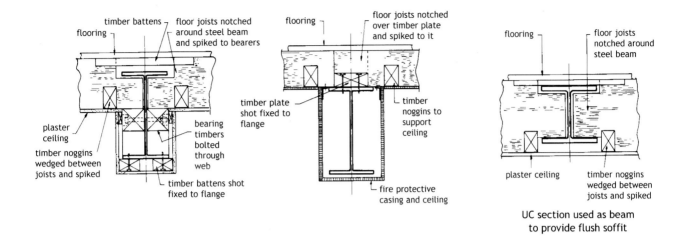

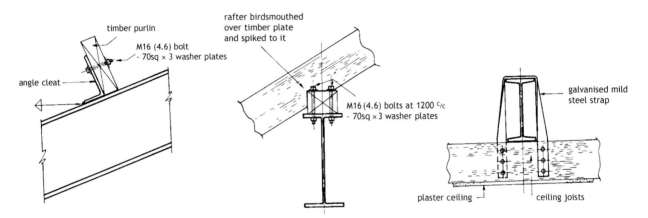

Figure 5.10 Typical steel/timber connections.

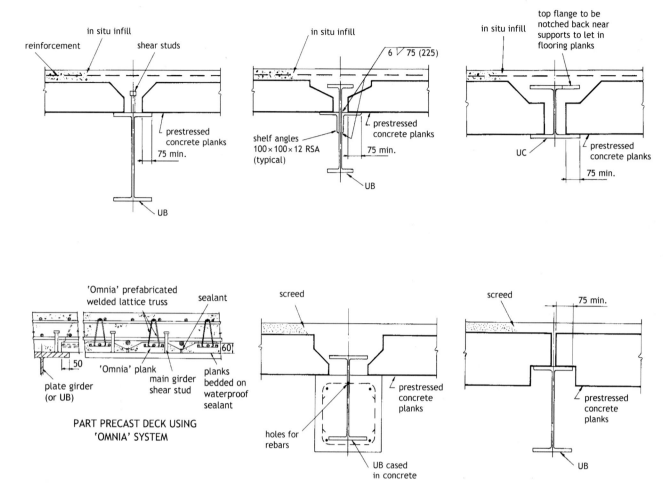

Figure 5.11 Typical steel/precast concrete connections.

6 Computer Aided Detailing

6.1 Introduction

Civil and structural engineers were one of the first groups to make use of computers. The ability to harness the computer's vast power of arithmetic made matrix methods of structural analysis a practical proposition. From this early beginning a whole range of computer programs and associated software has been developed to deal with most aspects of analysis and design. In the early days the use of the computer to produce drawings, while possible, did not receive much widespread attention. But now the use of computers for design and draughting can be said to have been the second industrial revolution.

Computer draughting systems have been available as commercial products since the 1970s. Most of the early systems were developed by the electronics industry to meet its own needs in the production of printed and integrated circuits. To the civil and structural engineer these early systems seemed little more than electronic tracing machines and of no great practical use. However, they formed the basis for the subsequent developments of systems more suited to construction.

The use of the computer to produce drawings differs in many ways from its use in analysis, design and other numeric activities, and computer draughting is substantially different from the traditional manual method.

The essential item of equipment now used is known as the workstation. Add-on peripherals might comprise plotters, including three-dimensional (3-D) plotters allowing rapid prototyping (refer to section 6.6), and scanners. While the input to and output from a draughting system are in graphical form, the computer's own representation of a drawing is as a mathematical model. This is a very important point as it is the nature of this 'model' that dictates the ease or difficulty with which different draughting systems perform what, to the end user, is the same drawing task. Since there is now a wide variety of specialist software available, users can become very knowledgeable, and this can result in a strong ability to transfer such skills although this may take considerable time.

In the early days much computer draughting development was undertaken by large companies who produced and maintained their own 'in-house' systems. Virtually no interaction could take place between these individual systems, principally due to the inconsistent computer language adopted by each company. Also most of these systems were driven by the company's mainframe computer which lacked sufficient memory, and because other software was used alongside (accounts, purchasing, etc.), the real time delays in carrying out work produced much frustration among staff.

With the evolution of the PC from a non-graphical low spec computer to the modern high-speed graphics workstation the power and the capabilities have developed to put very sophisticated tools in the hands of the detailer.

6.2 Steelwork detailing

It is a well known fact that structural steelwork is a highly complex three dimensional problem. Within a steel structure, connections will often comprise several intersecting members, originating from any number of different directions. The tasks of resolving such geometry into sound connection details and the production of fabrication drawings have always been extremely problematical. Traditionally, skilled draughtsmen with many years of detailing experience have been required.

Steel Detailers' Manual, Third Edition. Alan Hayward and Frank Weare. Third edition revised by Anthony Oakhill.
© 2011 Alan Hayward, Frank Weare and Anthony Oakhill. Published 2011 by Blackwell Publishing Ltd.

The constructional steelwork industry has continued to experience enormous economic and technological upheavals in recent years. In order to remain competitive, the majority of steelwork contractors have turned to new technologies in order to minimise their costs and meet the tighter deadlines which are being imposed by clients. After 2-D computer aided design (CAD) modelling the advent of 3-D parametric modelling of structural steelwork has proved beyond doubt to be one of the most viable solutions to the recent problems faced by steelwork fabricators.

A parametric feature-based modeller is a CAD software package that uses either a constructive solid geometry (CSG) or a boundary representation (B-REP) modeller that allows a user to refer to features instead of the underlying geometry. A feature is a term referring to higher order CAD entities. For example, given a 3-D splice plate with a bolt hole, the *hole* is considered a feature in the *plate* to reflect the manufacturing process used to create it, rather than referring to the hole mathematically as a cylinder. Parametric feature-based modellers use change states to maintain information about building the model and use expressions to constrain associations between the geometric entities. This ability allows a user to make a modification at any state and to regenerate the model's boundary representation based on these changes.

In building design, the principal means of communicating design intent is the drawing, whether it is a sketch, a concept design or a construction document. The traditional method of pen and drawing board requires skilled draughtsmen, who over the years have been in ever-decreasing supply. Each item is detailed independently and substantial checking is required to ensure that elements fit together. It is difficult to standardise details on a contract divided between several draughtsmen. All material lists, bolt lists and computer numerical control (CNC) programs must be produced manually by interpreting the detailed drawings. There are many potential sources for error.

The first CAD systems were effectively electronic drawing boards, allowing the user to create lines, circles, text and dimensions which duplicated the manual process, with the objective of creating the same drawing as before. In 2-D CAD, basic facilities such as move, copy, rotate, delete, etc. were introduced to speed up the process. Some 2-D CAD systems may have contained several parametric routines and libraries specifically for detailing steelwork. These would have assisted the manual detailing process and enabled better standardisation. However, each item was still

detailed independently and would have generally required the same substantial checking as manual draughting.

In the links between the designer and detailer, finite element analysis programs required the engineer to directly create a data file, which the analysis program could read. Most packages now have some sort of graphical input but are aimed specifically at creating analysis model data. 2-D CAD programs are then used to create the drawings that communicate this design intent to the steelwork fabricator. Engineers of course need software to enable them to model the steel structure for their own benefit, for analysis/design and integration with other disciplines. The fabricator can then use the resultant steel model with the detailed model returned to the engineer for checking and monitoring purposes. The relative ease of use and cost-effectiveness of 2-D systems means that they are still a valid solution, particularly for the creation of general arrangements drawings, especially in the design and build arena.

The 3-D modelling solution, on the other hand, is an entirely different concept from manual or 2-D CAD draughting. The steelwork structure is modelled in 3-D, rather than each item being drawn separately. The draughtsman does not in fact draw, instead he models. However, he is still a draughtsman, as the 3-D modelling system is his new tool and it will require his input and detailing knowledge.

The 3-D model, then, is a complete description of all steelwork, bolts, welds, etc. which constitutes all or part of a steel structure. It may contain any information whatsoever about any element within the structure. The steel structure actually exists, perfectly to scale, inside the computer. At any stage of the construction of the 3-D model, detailed drawings, listings or any other information may be produced completely automatically by the system. Once created, the database of information can be utilised by other parts of the software, to generate data in different ways such as detail drawings, general arrangements, materials lists, numerical control (NC) data, etc. The steelwork contractor knows that if the data (i.e. the model) is correct, then all the subsequent data will also be correct, so there is no need to check the drawings for dimensional accuracy. The 3-D model is the central source of all information, as shown in figure 6.1. A further goal is to export the same model to the design software. This is used by many companies, and in many instances this is the only way they work. Also, some modelling software now comes with analysis tools already built in.

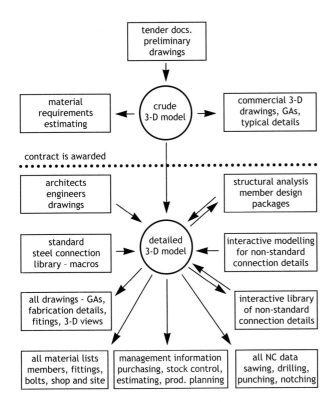

Figure 6.1 The central role of the 3-D modelling system.

6.3 Constructing a 3-D model of a steel structure

All steelwork structures are created within a 3-D framework of vertical grids and horizontal datum levels. The draughtsman will input these into the 3-D model, in accordance with the architect's or consulting engineer's general arrangement drawings.

The sizes of the principal members in a structure will generally have been determined by an engineer. In addition, member end reactions are often supplied to the fabricator for the design of connections. It is often the case that members will have been offset horizontally and/or vertically from grids and levels to meet architectural requirements.

The draughtsman will input members into the 3-D model, complete with correct sizes, offsets and end reactions (if supplied). Modern systems can model the member definition as well. This can have significant benefits with complicated setting-out problems. The definition of principal members will be extremely simple, in fact similar to drawing lines in 3-D. Initial member definition is done between set-out points and before connections are added.

Having established the geometric layout of the structural frame, the draughtsman must select the types of connec-

tions to use. The 3-D modelling system must have a comprehensive library of different connection types for the standard connections used in the construction of commercial and industrial buildings. In addition, the library may also include connections for the cold rolled products of major manufacturers. Figure 6.2 shows part of a typical connection library for a 3-D modelling system.

The connection library should allow the draughtsman to set up all the parameters for any connection type to suit both the company's and the client's standards and preferences. A single parametric set up for any connection type can then be applied to all kinds of different configurations and member sizes. The library should also be capable of designing a wide range of common connections (with associated calculation output) for the end reactions input by the draughtsman onto the 'wireframe' model.

It is considered essential by many that the 3-D modelling system should incorporate a powerful interactive modelling facility. The term 'interactive modelling' is used to describe the process of constructing a detail from first principles. This could also be used to modify and enhance an existing standard library connection. In addition to the creation of actual elements such as plates, bolts and welds, there is also the definition of the operations which are required to be carried out on the member, for example cutting a member to a plane (such as a rafter to the face of a stanchion) or cutting out parts of members to create openings or notches. The draughtsman must be able to easily create and modify any type of detail which it is possible to manufacture in the fabrication workshop. In addition, it must be possible to save interactively modelled details to a library, so that they may be reused on any particular contract.

The 3-D modelling system must allow automatic production of output at any stage of the model construction. There are generally two levels in this hierarchy. The first is Phase – this is the subdivision of a building or a contract; it could be a floor or the columns or an independent structure. The second is Lot – this is a further subdivision to facilitate planning of fabrication and delivery to site; it could be a lorry load or an erection group. Many steelwork contractors manufacture steelwork in phases which are linked to the erection programme. Very often the phase of steelwork is allied to the allowable limit carried on a transport lorry. It must therefore be possible to produce a 'phased' output of fabrication details, material lists and CNC data from the 3-D model. It should be noted that CNC is not specifically the direct link to the workshop machinery. In fact it is more a

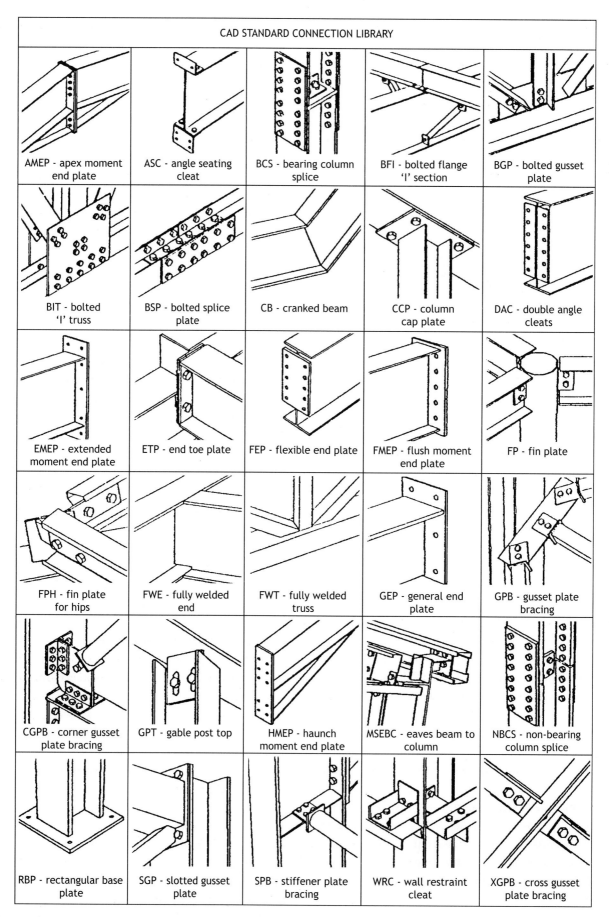

Figure 6.2 Typical standard steelwork connection library.

case of links to the NC machine software systems. DSTV has grown from being a German standard to become the de facto worldwide standard for the definition of geometry in NC systems for structural steelwork. DSTV is what most systems will now produce by default.

In summary then the 3-D modelling system should be capable of producing, and easily revising, all of the following different forms of output:

(1) Shop fabrication details

For all members, assemblies and fittings.

(2) Full size templates

For gusset plates and wrap-around templates for tubes.

(3) General arrangement drawings

Plans, elevations, sections, foundations, etc.

(4) Erection drawings

Realistic 3-D hidden views for any part of the structure.

(5) Materials lists

Cutting, assembly, parts, bolts, etc.

(6) CNC manufacturing data

Direct links to all types of workshop machinery.

(7) Interfaces to management information systems (MIS)

Purchasing, stock control, estimating, production management, accounting, databases, etc.

(8) Connection design calculations

For standard connections, in accordance with BS 5950 and UK industry accepted publications.

CNC sawing, cutting and drilling machines as well as robot welding machines will derive their instructions from information contained within a 3-D model. The entire management of steelwork design, manufacture and construction is now in the computerised hands of the MIS.

3-D modelling systems are now well established in the structural steelwork industry. Fabricators can already place orders with their suppliers through MIS links from their 3-D systems. The design and detailing of steel structures has become more integrated, with consulting engineers and design offices imparting information to fabricators electronically, instead of providing general arrangement drawings. However, where a 3-D model has been created in an engineer's office it generally will exist in some other software model. This will require the transfer of 3-D steel information between different systems. Many software applications can now accept and export a wide range of formats.

In recent years CIMsteel Integration Standards CIS/1 and now CIS/2 have been developed to provide a means of transferring complete building model information between the various types of system employed in the industry. The CIS are a set of information specifications. They provide standards against which the vendors of engineering application software can develop and implement translators. These translators enable the users of such software to export engineering data from one application and import into another. Thus, the CIS (developed from the Eureka CIMsteel Project) can be used to transfer 'product data' (information about a specific steel frame) between applications software packages, whether they are located within the same company or in different companies.

6.4 Object orientation

Traditional CAD systems, such as AutoCAD, are now not simply methods of creating lines and text on a drawing. They are becoming platforms to enable software applications to model and manipulate 'objects' in an intelligent way. The concept of 'object modelling' is that the definition of an object is contained within the object itself upon creation. Obviously, the software that created the object in the first place understands what it is and what the data mean. The idea is that different software packages can access the object and deal with the different aspects of the data as required.

For instance the various elements of a steel modelling system will understand the concepts of what a piece of steel is, the meaning of a section size, the relevance of a bending moment and connection design forces. If one piece of steel clashes with another, say a beam and a column, or if something changes, then the system has rules or 'methods' to determine what action to take. By creating the model from real components such as beams, columns, slabs, etc. on to which the engineer can apply loading and constraints, and by further defining the type of connectivity, the system will determine the appropriate degree of restraint. This will eventually be taken into account when the element and connection design is carried out.

6.5 CNC/rapid prototyping

One exciting new development has been the introduction of CNC/rapid prototyping (RP). These are a range of technologies that cut or build physical objects direct from computer CAD files. CNC/RP has been developed in an industry context and over the past few years its use by engineers, architects and artists has increased.

There are two basic groups, each with a range of processes:

(1) *Building (rapid prototyping),* which builds 3-D objects in a range of materials using a system that converts computer-generated designs into a series of very fine layers or slices.
(2) *Cutting and milling,* which cuts or shapes existing materials, such as timber, plastic or metal. Cutting is generally applied to materials in sheet form while milling generally involves shaping an object on a lathe.

Rapid prototyping takes virtual designs from CAD and transforms them into thin, virtual, horizontal cross-sections and then creates each cross-section in physical space, one after the next until the model is finished. It is a WYSIWYG process where the virtual model and the physical model correspond almost identically.

With additive fabrication, the machine reads in data from a CAD drawing and lays down successive layers of liquid, powder or sheet material, and in this way builds up the model from a series of cross-sections. These layers, which correspond to the virtual cross-section from the CAD model, are joined together or fused automatically to create the final shape. The primary advantage of additive fabrication is its ability to create almost any shape or geometric feature.

The standard data interface between CAD software and the machines is the stereolithography (STL) file format. An STL file approximates the shape of a part or assembly using triangular facets. Smaller facets produce a higher quality surface. The word 'rapid' is relative: construction of a model with contemporary methods can take from several hours to several days, depending on the method used and the size and complexity of the model. Additive systems for rapid prototyping can typically produce models in a few hours, although this can vary widely depending on the type of machine being used and the size and number of models being produced simultaneously. Figure 6.3 shows a typical flow guide summarising CNC/rapid prototyping.

Rapid prototyping is now entering the field of rapid manufacturing and it is believed by many experts that this is a 'next level' technology.

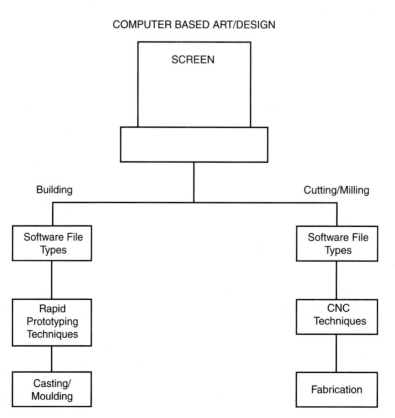

Figure 6.3 CNC/rapid prototyping guide.

6.6 Future developments

The widespread adoption of CAD by all sectors of the constructional steelwork industry has enabled drawings to be sent electronically from one office to any other office. The CAD drawing is read into another system, using any one of a number of formats, to be used as a basis for subsequent drawings. This can give rise to the question of responsibility for data integrity, since it is still possible to create a CAD drawing incorrectly. Currently, it is the norm that paper representation of the CAD drawing and its interpretation are probably viewed as more valid than an electronic version. Generally, at present, if the engineer wishes to give approval to the fabricator's work then the only way is still from the detail drawings, since there is no way of using the data in the fabricator's model. Similarly, if the steelwork contractor wants to issue information to a sub-contractor then it will be issued as paper drawings, or at best as CAD files.

Previously, a 3-D data exchange file model imported into a 3-D steel modelling system generally had no use. The only benefit was that it could be used as a background image to which objects could be snapped. Ideally what was needed was the intelligent transfer of data between systems, whether that information was be based on analysis, design or detailing. The preferred solution here rests with the continued successful adoption by the industry of CIS product developments.

When the model is passed to others in the design chain, then the data includes not only the sizes and positions of members but also the forces, connection design assumptions and any other necessary information. This is the basis for co-operative working in a quality assured environment. The proliferation of the internet has provided an overpowering means for communicating and sharing data. Whereas in the past the data was passed from one company to another, nowadays data is stored centrally and regularly accessed by each member of the design team.

There are still many problems with this flow of information which ultimately waste time and money for all those concerned. Better use of software technology and applications should in the long term be able to improve this situation. Those working in structural steelwork have for some time had a wide range of software tools to assist them. There is, however, a new way of working emerging which involves an integrated approach with the steelwork supply chain and other disciplines working together to generate full building models in 3-D. Steelwork detailers are well advanced in their use of models but there is a whole range of tools needed in other parts of the supply chain. These involve both the data standards to permit the sharing and transfer of information together with the development of the objects to take full advantage of the opportunities which can be derived from the emerging technology.

There are a number of other applications also available that allow a user to import a number of model formats into one common space, and to review all aspects of the works and perform clash detection.

Much has been written about the 'paperless office', and there is a variety of software that allows the user to review a drawing on screen and 'red line' corrections and comments. The originator of the drawing can then open the drawing, review the comments as a markup and proceed to incorporate the required changes, without the need to produce any paperwork.

The increasing sophistication of the software now available allows the industry to undertake much more spectacular detailed designs. If a free-form organic model is taken that can be re-configured to become an architectural form, then a rationalised structural frame can be applied to it with ease. Then the interaction between the software and the CNC workshop machines makes the seemingly complex fabrication possible.

One of the latest developments is the single model environment, which is now being used by many designers and detailers. Basically, everyone associated with a project uses the same model to ensure there are no fit problems. All disciplines on the project are co-ordinating off the same information. This generally requires an extranet site for the models to be loaded onto, and all parties must use similar software packages.

7 Examples of Structures

Following are examples of various types of structures utilising structural steelwork. Some of these are taken from actually constructed projects designed by the authors. The practices and details shown will be suitable for many countries of the world. The member sizes are as actually used where shown, but it is emphasised that they might not always be appropriate in a particular case, because of variations in loading or requirements of different design codes.

A brief description of each structure type is included, giving particular reasons for use and any particular influences which affect the method of construction or details employed.

7.1 Multi-storey frame buildings

Multi-storey steel frames provide the structural skeleton from which many commercial and office buildings are supported. Steel has the advantage of being speedy to erect and it is very suitable in urban situations where conditions are restrictive. This is further exploited by the use of rapidly constructed floors and claddings. This means that a 'dry envelope' is available at the earliest possible date so that interior finishes can be advanced and the building occupied sooner. Floor systems used include precast concrete and composite profiled galvanised metal decking, which can also be made composite with the steel frame. Such decking is supplied in lengths which span over several secondary beams and shear studs are then welded through it. Mesh reinforcement is provided to prevent cracking of the concrete slab.

The structural layout of beams and columns will largely depend upon the required use. Modern buildings require extensive services to be accommodated within floors and this may dictate that beams contain openings. Here castellated or tapered beams can be useful. In general, floors are supported by secondary and main beams usually of universal beams, supported by columns formed from UCs. The spacing of secondary beams is dictated by the floor type, typically 2.5 m to 3.5 m. An important design decision is whether stability against horizontal forces (e.g. due to wind or earthquake) is to be resisted using rigid connections or whether bracing is to be supplied and simple connections used. Alternatively, other elements may be available such as lift shafts or shear walls, allied with the lateral rigidity of floors, to which the steelwork can be secured. In this case temporary stability may need to be supplied using diagonal bracings during erection until a means of permanent stability is provided.

The example shown in figures 7.1 to 7.5 is a two-storey office building with floors and roof of composite profiled steel decking. Beam to column connections are of simple type, and stability is provided by wind bracings installed within certain external walls. Because there are only two storeys the columns are fabricated full height without splices. The top of the columns can be detailed to suit future upward extension if required. Connections for the cantilevered canopy beams are of rigid end plate type.

Figure 7.2 is a first floor part plan, being part of the engineer's drawings, which gives member sizes and ultimate limit state beam reactions for the fabricator to design the connections. Typical connections are shown in figure 7.3. Workshop drawings of a beam and a column are shown in figures 7.4 and 7.5 respectively which are prepared by the fabricator after designing the connections.

Steel Detailers' Manual, Third Edition. Alan Hayward and Frank Weare. Third edition revised by Anthony Oakhill.
© 2011 Alan Hayward, Frank Weare and Anthony Oakhill. Published 2011 by Blackwell Publishing Ltd.

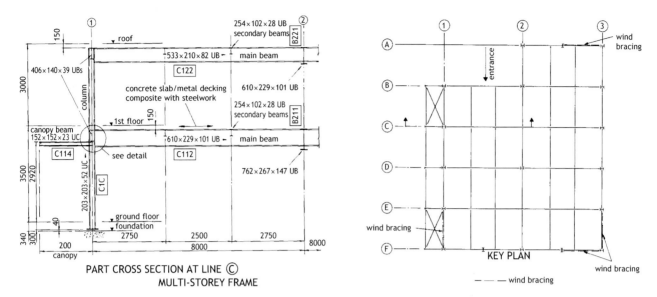

Figure 7.1 Multi-storey frame building.

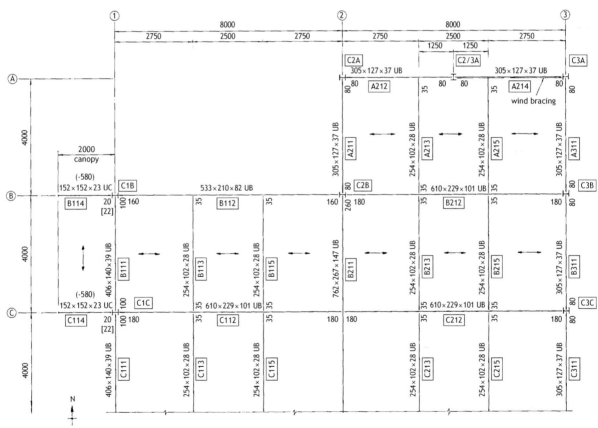

GENERAL NOTES

1. Reactions are factored loads to BS 5950 in kN. Bending moments if any in kNm in brackets e.g. [180].
2. All steel to EN 10025 grade S275.
3. All beam marks to be at north or east end. All column marks to be on flange facing north or east.
4. All beams at 150 below 1st floor except where shown in brackets e.g. (−580).
5. ⟷ indicates the direction of metal decking or cladding.

**STEELWORK
FIRST FLOOR PLAN**

Figure 7.2 Multi-storey frame building.

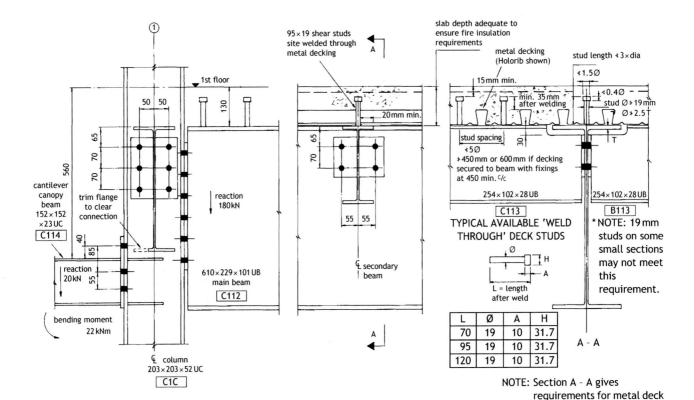

TYPICAL CONNECTIONS AT FIRST FLOOR
GRID REF ① - ©

Figure 7.3   Multi-storey frame building.

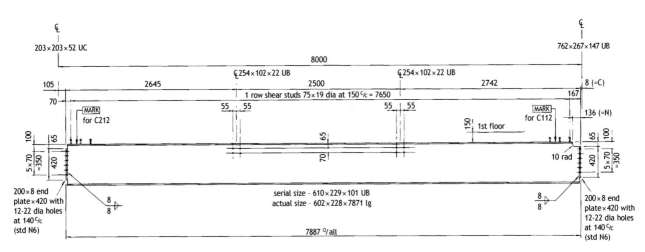

1-BEAM REQ'D AS DRAWN & NOTED MK'D C112
1-BEAM REQ'D AS DRAWN & NOTED MK'D C212

GENERAL NOTES
Unless otherwise noted
1. All material to EN 10025 grade S275
2. All bolts M20 grade 4.6
3. All holes 22 dia
4. Treatment — see spec.

WORKSHOP DRAWING — BEAM DETAIL

Figure 7.4   Multi-storey frame building.

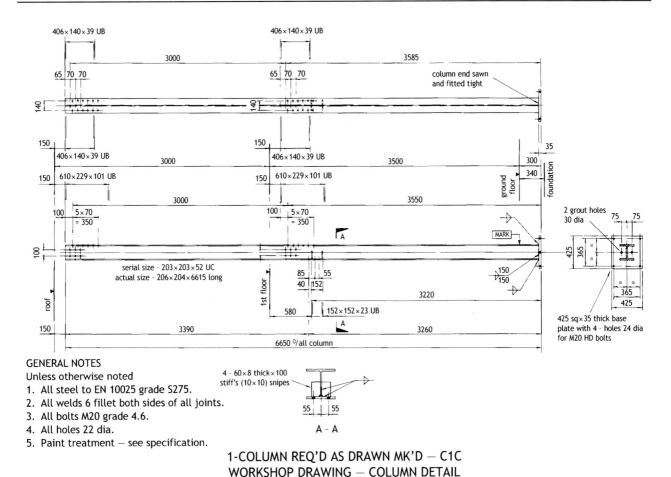

GENERAL NOTES

Unless otherwise noted

1. All steel to EN 10025 grade S275.
2. All welds 6 fillet both sides of all joints.
3. All bolts M20 grade 4.6.
4. All holes 22 dia.
5. Paint treatment – see specification.

1-COLUMN REQ'D AS DRAWN MK'D – C1C
WORKSHOP DRAWING – COLUMN DETAIL

Figure 7.5 Multi-storey frame building.

7.1.1 Fire resistance

Generally, multi-storey steel framed buildings are required by Building Regulations to exhibit a degree of fire resistance that is dependent on the building form and size. Fire resistance is specified as a period of time, e.g. $^1/_2$ hour, 1 hour, 2 hours, etc., and is normally achieved by insulation in the form of cladding. The thickness of cladding required is therefore dependent on material type and period of resistance. Traditional materials such as concrete, brickwork and plasterboard are still used but have to a great extent been replaced by modern lightweight materials such as vermiculite and mineral fibre. Asbestos is no longer used for health reasons.

Lightweight claddings are available in spray form or board; sprays, being unsightly, are generally used where they will not be seen, e.g. floor beams behind suspended ceilings. Boards can be prefinished or decorated and are fixed typically by screwing mainly to noggins or wrap-around steel straps. Typical arrangements are shown in figure 7.6. The thickness of cladding and fixing clearly affects building details and therefore warrants early consideration.[7]

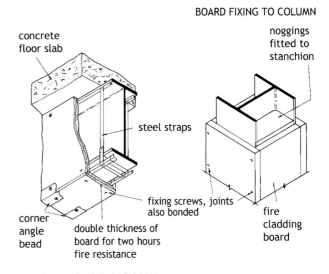

BOARD FIXING TO FLOOR BEAM

Figure 7.6 Multi-storey frame building.

7.2   Single-storey frame buildings

Single-storey frame buildings are extensively used for in-dustrial, commercial and leisure buildings. In many coun-tries of the world they are economically constructed in steel because the principal loads, namely the roof and wind, are relatively light, yet the spans may be large, commonly up to about 45 m. Steel with its high strength : weight character-istics is ideally suited for single-storey buildings. The frame efficiently carries the roof cladding independently of the walls thus offering flexibility in location of openings or partitions. Side cladding is directly attached to the frame which gives stability to the whole building. This system is also ideally suited to structures in seismic areas. Sometimes solid side cladding such as brickwork is used part or full height, and it is often convenient to stabilise this by attach-ment to the frame although vertical support is independent.

Generally the steel frame terminates at least 300 mm below floor level on its own foundations. This permits flexibility in future use of the floor, which may need to contain openings or basements and be replaced periodically if subjected to heavy use. Any internal walls or partitions are generally not structurally connected to the frame so that there is flexibil-ity in relocation for any different future occupancy.

Figure 7.7 shows a number of frame types. A single bay is indicated but multiple bays are often used for large buildings for economy when internal columns are permitted. Portal frames, the most common type, are described in section 7.3.

Requirements for natural lighting by provision of translucent sheeting or glazing often govern roof shape and therefore the type of frame. In particular the monitor roof type (figure 7.7(j)) provides a high degree of natural light. The

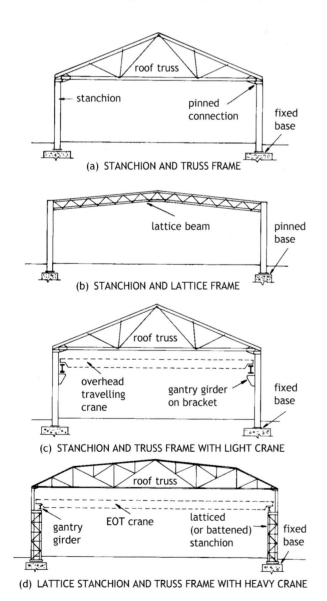

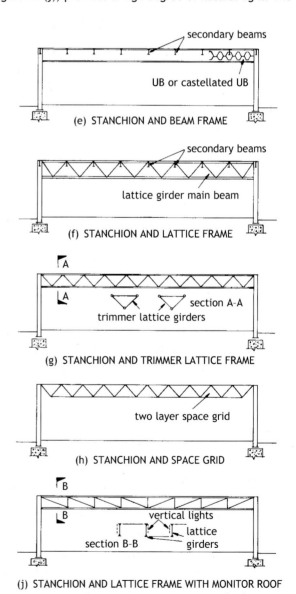

Figure 7.7   Single-storey frame building.

widespread use of lightweight claddings, especially profiled steel sheeting (usually galvanised and plastic coated in a range of colours), which have largely displaced other materials, permits economic roofs of shallow pitch (typically 1 : 10 or 6°). Such cladding is available with an insulation layer, which can, if necessary, be incorporated below purlin level to produce a flush interior if needed for hygienic reasons. Flat roofs, but with provision for drainage falls, covered by proprietary roof decking are also used, but at generally greater expense. Sufficient camber or crossfall must be used to ensure rainwater run-off. Depending upon the required use, provision of a suspended ceiling may also decide the frame type. For industrial buildings internal cranes are usually required in the form of electric overhead travelling (EOT) type supported by gantry girders mounted on the frame. Clearances and wheel loads for the crane (or cranes) must be considered, which will vary according to the particular manufacturer.

The structural form most generally used is the portal frame described in section 7.3. Figure 7.7 shows a number of other types. The stanchions and truss type frames (a) and (c) are more suited to roofs having pitch greater than 3 : 10. Presence of the bottom tie is convenient for support of any suspended ceilings, but a disadvantage is that the stanchion bases must be fixed to ensure lateral stability. The lattice stanchion and truss frame (d) is suitable for EOT cranes exceeding 10 tonnes capacity. Where appearance of the frame is important or where industrial processes demand clean conditions, hollow section members are suitable using triangular lattice girders as (g) or space grids (h). The latter are uneconomic for spans up to about 40 m, but are suitable for long spans if internal stanchions are not permitted.

Bolted site connections are generally necessary between stanchions and roof structure with the latter fabricated full span length where delivery allows. Truss or lattice roofs usually have welded workshop connections. Secondary members in the form of sheeting rails or purlins are usually of cold formed sections (see section 7.3). A vital consideration is longitudinal stability, especially during erection, which requires the provision of bracing to walls taking account of the location of side openings. Roof bracing is also necessary except where plan rigidity is inherent such as with a space grid. Gantry girders for EOT cranes should incorporate details which permit adjustment to final position as shown in figure 7.8, and possible replacement of rails during the life of the structure. Safety requirements such as space for personnel between end of crane and structures and positioning of power cables must be met.

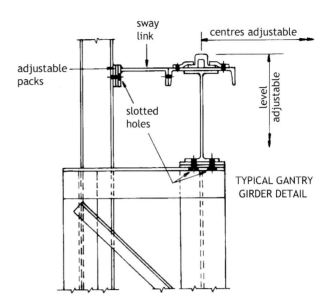

Figure 7.8 Single-storey frame building.

7.3 Portal frame buildings

Steel portal frames (Figures 7.9 and 7.10) are the most common and are a particular form of single-storey construction. They became popular from the 1950s and are particularly efficient in steel, being able to make use of the plastic method of rigid design which enables sections of minimum weight to be used. Frame spacings of 4.5 m, 6.0 m and 7.5 m with roof pitch typically 1 : 10, 2 : 10 and 3 : 10 are common. Portal frames provide large clear floor areas offering maximum adaptability of the space inside the building. They are easily capable of being extended in the future and, if known at the design stage, built-in provision can be made. Multiple bays are possible. Variable eaves heights and spans can be achieved in the same building and selected internal columns can be deleted where required by the use of valley beams. Portal frames can be designed to accommodate overhead travelling cranes typically up to 10 tonnes capacity without use of compound stanchions.

Normally, wind loads on the gable ends are transferred via roof and side bracing systems within the end bays of the building to the foundations. The gable stanchions also provide fixings for the gable sheeting rails, which in turn support the cladding. Cold rolled section sheeting rails and purlins are usual, but alternatively hot rolled steel angle sections are suitable. Various proprietary systems are available using channel or zed sections. The sleeved system is popular whereby purlins extend over one bay between portal frames, but are made continuous over intermediate portals by a short sleeve of similar section. The systems often offer a

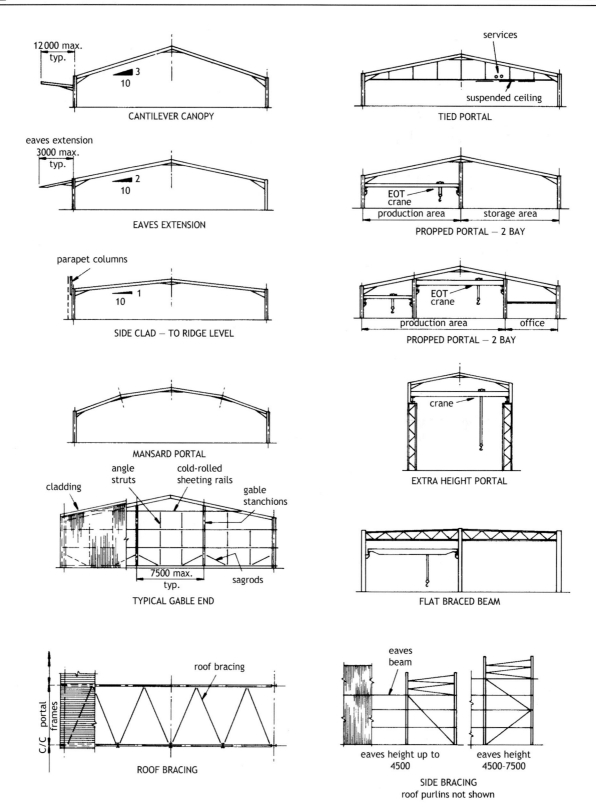

Figure 7.9 Portal frame buildings.

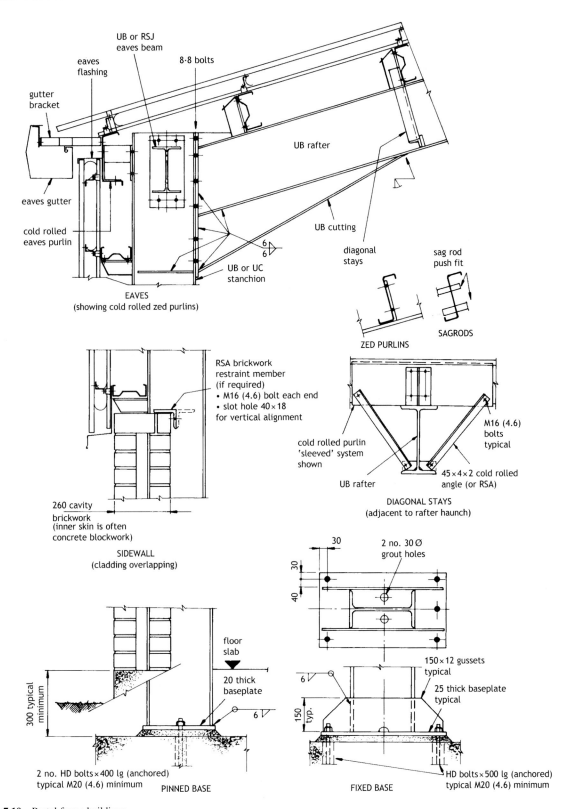

Figure 7.10 Portal frame buildings.

Span of purlins	No. rows
⊅ 4500	—
> 4500 to 7600	1
> 7600 to 10000	2
Sag rods	

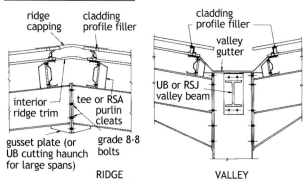

RIDGE   VALLEY

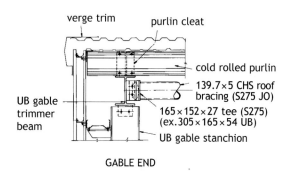

GABLE END

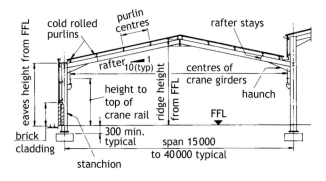

TYPICAL PORTAL FRAME

Figure 7.10   *Contd*

range of fitments including rafter cleats, sag rods, rafter restraints, eaves beams, etc.

Main frame members are normally of universal beams with universal columns sometimes being used for the stanchions only. Tapered haunches (formed from cuttings of rafter section) are often introduced to strengthen the rafters at eaves, especially where a plastic design analysis has been used. Either pinned or fixed bases may be used. Main frames of tapering fabricated section are used by some fabricators,

some of whom offer their own ranges of standard portal designs.

Bracing is essential for the overall stability of the structure especially during erection. Different arrangements from those illustrated may be necessary to accommodate door or window openings. It is important to provide restraint against buckling of rafters in the eaves region, this usually being supplied by an eaves beam together with diagonal stays connected to the purlins. Wind uplift forces often exceed the dead weight of portal frame buildings due to low roof pitch and light weight, such that holding down bolts must be supplied with bottom anchorage. Reversal of bending moments may also occur at eaves connections.

7.4   Vessel support structure

The structure (figures 7.11, 7.12, 7.13 and 7.14) supports a carbon dioxide vessel weighing 12 tonnes and 1.9 m diameter × 5.2 m long, approximately 3.1 m above ground level. It is typical of small supporting steelwork within industrial complexes and was installed inside a building. It comprises a main frame with four columns and beams made as one welded fabrication with rigid connections supporting the vessel cradle supplied by others. Access platforms are provided at two levels below and above the vessel with hooped access ladders.

Drawing notes

(1)   All steel to be EN 10025 grade S275 UON.
(2)   All bolts to be black bolts grade 4.6. To be M16 diameter UON.
(3)   All welds to be fillet welds size 5 mm UON continuous on both sides of all joints.
(4)   Protective treatment all at workshop:
       Grit blast 2nd quality and zinc-rich epoxy prefabrication primer.
       2 coats zinc-rich epoxy paint after fabrication.
       Total nominal dry film thickness 150 microns.

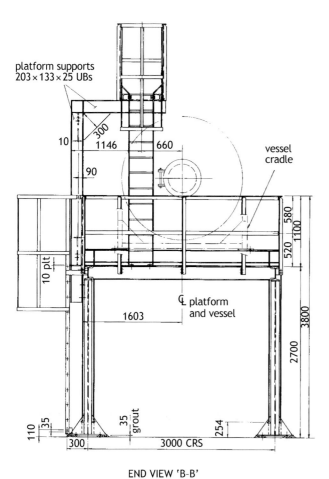

Figure 7.11 Vessel support structure.

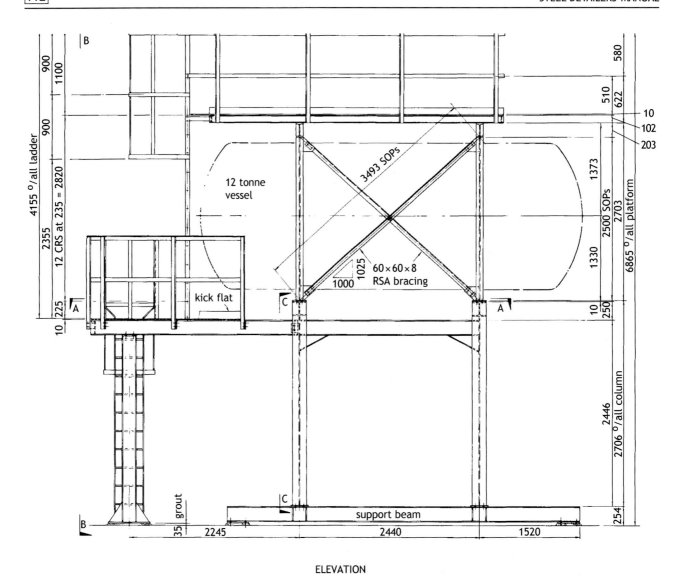

ELEVATION

Figure 7.12   Vessel support structure.

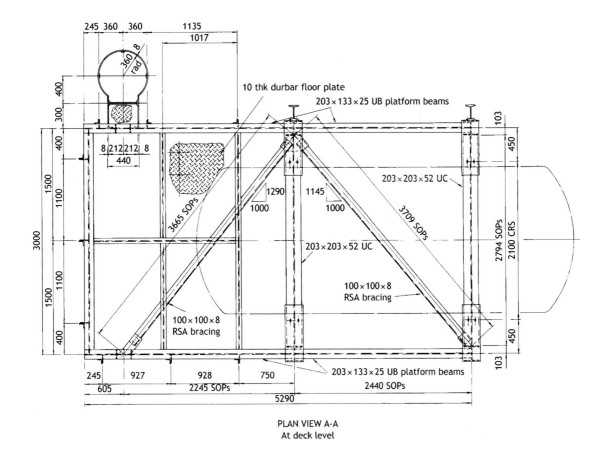

PLAN VIEW A-A
At deck level

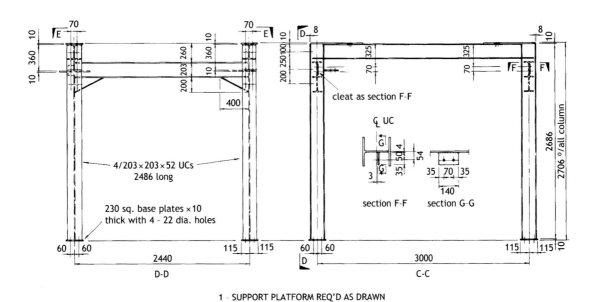

1 - SUPPORT PLATFORM REQ'D AS DRAWN

Figure 7.13 Vessel support structure.

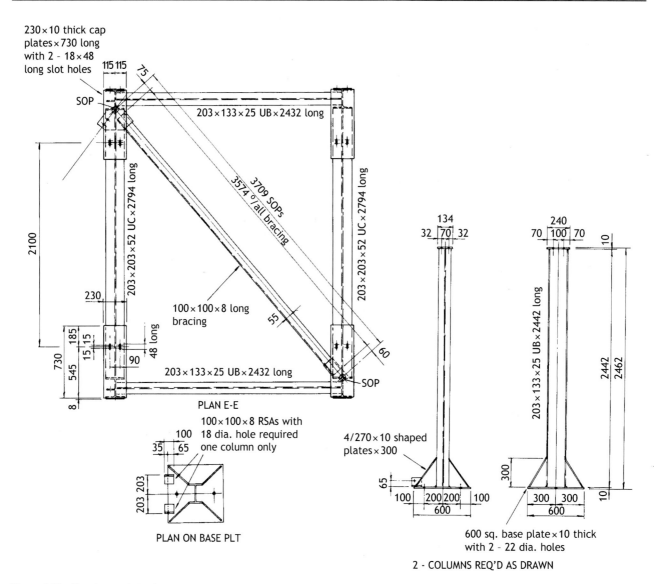

Figure 7.14 Vessel support structure.

7.5 Roof over reservoir

The roof (figures 7.15 and 7.16) provides a protective covering over a fresh water reservoir with a span of about 19.5 m which is clad with profiled steel sheeting. It comprises pitched universal beam rafters which are tied at eaves level with RSA ties because the reservoir edge walls are not capable of resisting outward horizontal thrust. The ties are supported from the ridge at mid-length to prevent sagging. Roof plan bracing is supplied within one internal bay to ensure longitudinal stability of the roof.

Drawing Notes

(1) All steel to be EN 10025 grade S275 UON.

(2) All bolts to be black bolts grade 4.6 UON.
   To be M16 diameter UON.

(3) All welds to be fillet welds size 6 mm UON continuous on both sides of all joints.

(4) Protective treatment:
   Grit blast 2nd quality and zinc-rich epoxy prefabrication primer.
   One coat zinc-rich epoxy paint at workshop.
   One coat zinc-rich epoxy paint at site after erection.
   Total nominal dry film thickness 150 microns.

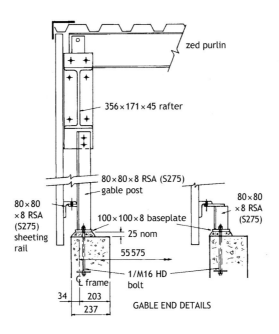

GABLE END DETAILS

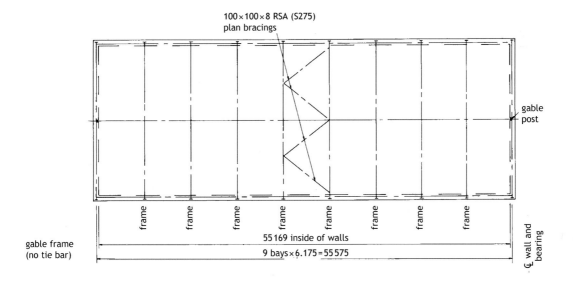

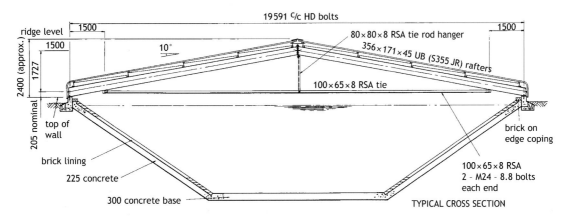

Figure 7.15 Roof over reservoir.

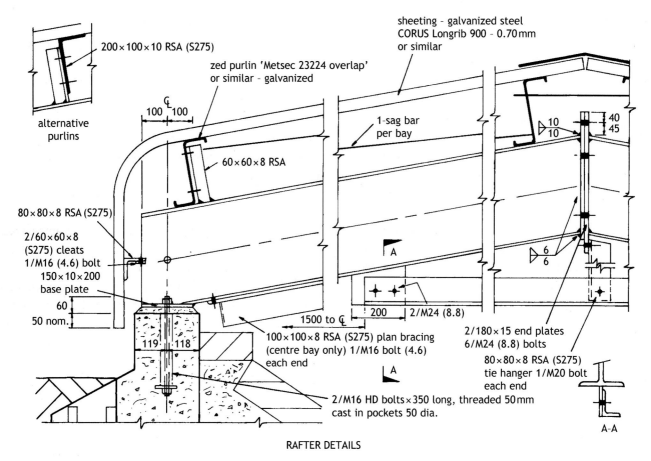

sheeting – galvanized steel
CORUS Longrib 900 – 0.70 mm
or similar

200×100×10 RSA (S275)

zed purlin 'Metsec 23224 overlap'
or similar – galvanized

alternative
purlins

100 100

60×60×8 RSA

1-sag bar
per bay

40
45
10
10

80×80×8 RSA (S275)

2/60×60×8
(S275) cleats
1/M16 (4.6) bolt

A

6
6

150×10×200
base plate

60
50 nom.

119 118

200   2/M24 (8.8)

1500 to ℄

100×100×8 RSA (S275) plan bracing
(centre bay only) 1/M16 bolt (4.6)
each end

A

2/180×15 end plates
6/M24 (8.8) bolts

80×80×8 RSA (S275)
tie hanger 1/M20 bolt
each end

2/M16 HD bolts×350 long, threaded 50 mm
cast in pockets 50 dia.

A-A

RAFTER DETAILS

Figure 7.16   Roof over reservoir.

7.6   Tower

The tower (figures 7.17, 7.18 and 7.19) is 55 m high and supports electrical equipment within an electricity power-generating station in India. It was fabricated in the UK and transported piecemeal by ship in containers. The major consideration in the design of tower structures is wind loading due to the height above ground and comparatively light weight of the equipment carried. Open braced structures are usual for towers so as to offer minimal wind resistance. Either hollow sections or rolled angles would have been suitable and although the former have an advantage in providing for smooth air flow and thus less wind resistance, the latter were chosen to simplify the connections. Use of bolted connections using gusset plates meant that all members could be economically fabricated using NC saw/drilling equipment.

(1)   All steel to be to EN 10025 grade S275 UON.
(2)   All bolts to be grade 4.6. To be M24 diameter UON.

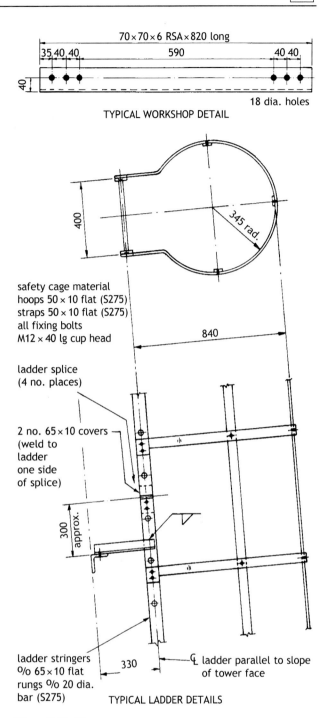

Figure 7.17   Tower.

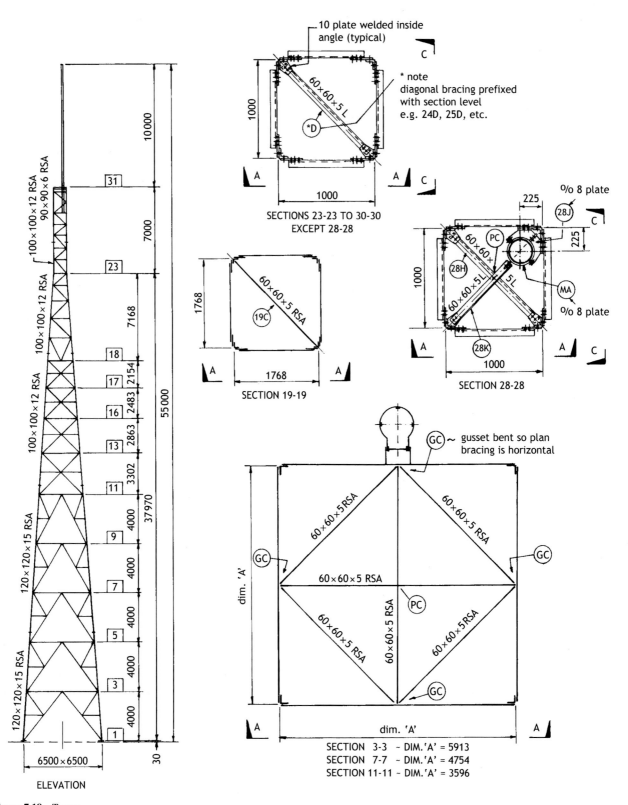

Figure 7.18 Tower.

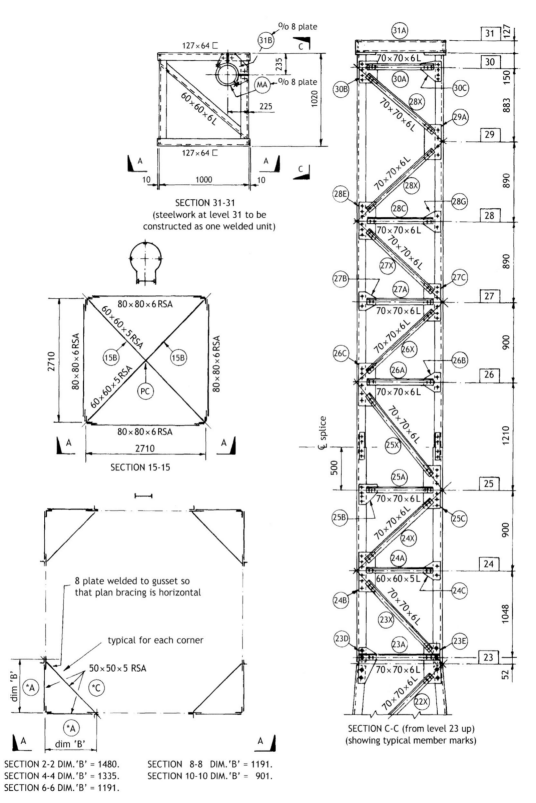

SECTION 31-31
(steelwork at level 31 to be
constructed as one welded unit)

SECTION 15-15

8 plate welded to gusset so
that plan bracing is horizontal

typical for each corner

50 × 50 × 5 RSA

SECTION 2-2 DIM.'B' = 1480.   SECTION 8-8 DIM.'B' = 1191.
SECTION 4-4 DIM.'B' = 1335.   SECTION 10-10 DIM.'B' = 901.
SECTION 6-6 DIM.'B' = 1191.

SECTION C-C (from level 23 up)
(showing typical member marks)

Figure 7.18 *Contd*

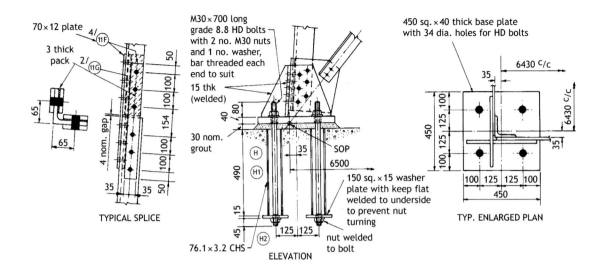

70 × 12 plate
3 thick pack
4 nom. gap
65
65
35   35
50
100 100 100 100
154
50
TYPICAL SPLICE

M30 × 700 long grade 8.8 HD bolts with 2 no. M30 nuts and 1 no. washer, bar threaded each end to suit
15 thk (welded)
30 nom. grout
490
15
45
76.1 × 3.2 CHS
150 sq. × 15 washer plate with keep flat welded to underside to prevent nut turning
nut welded to bolt
SOP
6500
35
125 125
ELEVATION

450 sq. × 40 thick base plate with 34 dia. holes for HD bolts
6430 c/c
6430 c/c
35
35
450
450
100 125 125 100
100 125 125 100
TYP. ENLARGED PLAN

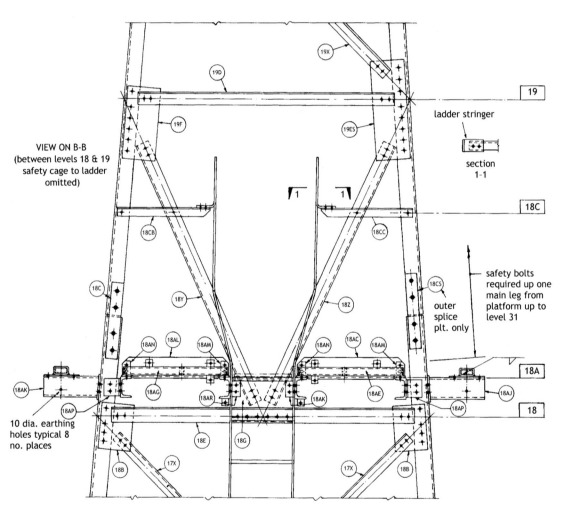

VIEW ON B-B
(between levels 18 & 19 safety cage to ladder omitted)

ladder stringer
section 1-1

safety bolts required up one main leg from platform up to level 31

outer splice plt. only

10 dia. earthing holes typical 8 no. places

Figure 7.19   Tower.

7.7 Bridges

Several developments in recent years have improved the status and opportunities for steel in bridges, increasing its market share over concrete structures in a number of countries.

Developments include:

(1) Fabricators have improved their efficiency by use of automation.
(2) Stability of steel prices with wider availability in many countries by opening of steel plants.
(3) Use of mobile cranes to erect large pre-assembled components quickly, thus reducing number of mid-air joints.
(4) Composite construction economises in materials.
(5) Permanent formwork or precasting for slabs.
(6) Improved protection systems using fewer paint coats having longer life.
(7) Use of unpainted weathering steel for inaccessible bridges.
(8) Use of site welded or HSFG bolted joints to achieve continuous spans.
(9) Better education in steel design.

For multiple short (up to 30 m) and medium (30–150 m) spans continuity is common with welded or HSFG bolted site joints to the main members. Articulation between deck and substructures is generally provided using sliding or pinned bearings mounted on vertical piers often of concrete but occasionally steel. Constant depth main girders are usual, with fabricated precamber to counteract deflection. Curved soffits are sometimes used (as shown in figure 7.20).

Curved bridges are often formed using straight fabricated chords with change of direction at site splices. Composite *deck type* cross sections are usual for highway bridges as shown in figure 7.21 and suit the width of modern roads except where construction depth is very restricted when half-through girders are used, especially for railway bridges as shown in figure 7.22. Multiple rolled sections are used for short spans with plate girders being used when the span exceeds about 25–30 m. Intermediate lateral bracings are provided for stability. Sometimes they are proportioned to assist in transverse distribution of live load, but practices vary between different countries. Box girders as shown in figure 7.22 are also used and open top boxes 'bathtubs' are extensively used in North America. Problems can arise during construction due to distortion and twisting of open top boxes prior to the rigidifying effect of the concrete slab being realised and temporary bracings are thus essential.

Most early composite bridges used *in situ* slabs cast on removable formwork supported from the steelwork. For many years the high costs of timber and site labour have encouraged permanent formwork. Various types are in use including profiled steel sheeting (especially in the USA), glass reinforced plastic (grp), glass reinforced concrete (grc) and part depth concrete planks. The 'OMNIA' type of precast unit is being used (see figure 5.11), which incorporates a welded lattice truss to provide temporary capacity to span up to about 3.5 m between steel flanges, whose lower chord is cast in. Extra reinforcement is incorporated supplemented by further continuous rebars at the 'vee' joints to resist live loads. Detailing of the slab needs to be carefully done to avoid congestion of reinforcement and allow proper compaction of concrete.

For *footbridges* steel provides a good solution because the entire cross section, including parapets, can be erected in one piece. Cross sections are shown in figure 7.23. Economic solutions use half-through lattice or Vierendeel girders with members of rolled hollow section and deck plate with factory applied epoxy-type non-slip surfacing 6 mm or less in thickness. Columns, staircases and ramps are also commonly of steel using hollow sections. For urban areas the half-through section achieves minimum length approach stairs or ramps. Further space can be saved by using *stepped ramps* which achieve an average slope of 1 in 6 compared with 1 in 10 for sloping ramps.

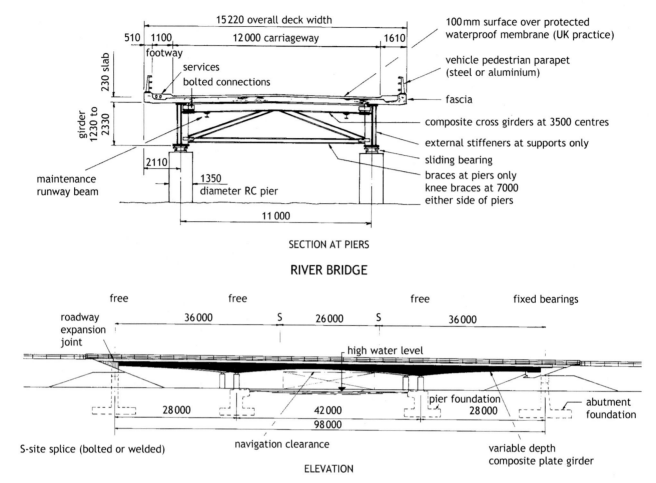

15 220 overall deck width

510 1100   12 000 carriageway   1610

footway

230 slab

services
bolted connections

girder
1230 to
2330

2110

maintenance
runway beam

1350
diameter RC pier

11 000

100 mm surface over protected
waterproof membrane (UK practice)

vehicle pedestrian parapet
(steel or aluminium)

fascia

composite cross girders at 3500 centres

external stiffeners at supports only

sliding bearing

braces at piers only
knee braces at 7000
either side of piers

SECTION AT PIERS

RIVER BRIDGE

free   free   free   fixed bearings

roadway
expansion
joint

36 000   S   26 000   S   36 000

high water level

28 000   42 000   28 000

pier foundation

abutment
foundation

98 000

S-site splice (bolted or welded)

navigation clearance

variable depth
composite plate girder

ELEVATION

Figure 7.20 Bridges.

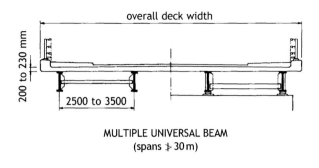

MULTIPLE UNIVERSAL BEAM
(spans ⊁ 30 m)

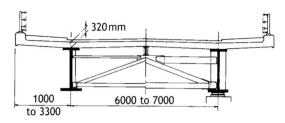

TWIN PLATE GIRDER & STRINGER
(spans 40 to 100 m

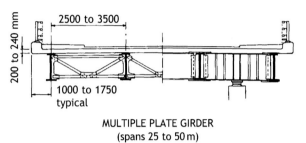

MULTIPLE PLATE GIRDER
(spans 25 to 50 m)

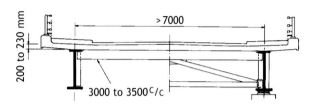

TWIN PLATE GIRDER & CROSS GIRDERS
(spans 40 to 100 m)

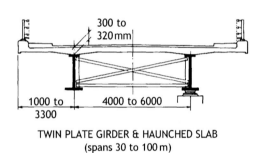

TWIN PLATE GIRDER & HAUNCHED SLAB
(spans 30 to 100 m)

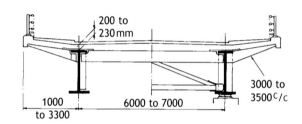

TWIN PLATE GIRDER & STEEL CANTILEVERS
(spans 40 to 100 m)

Figure 7.21 Bridges.

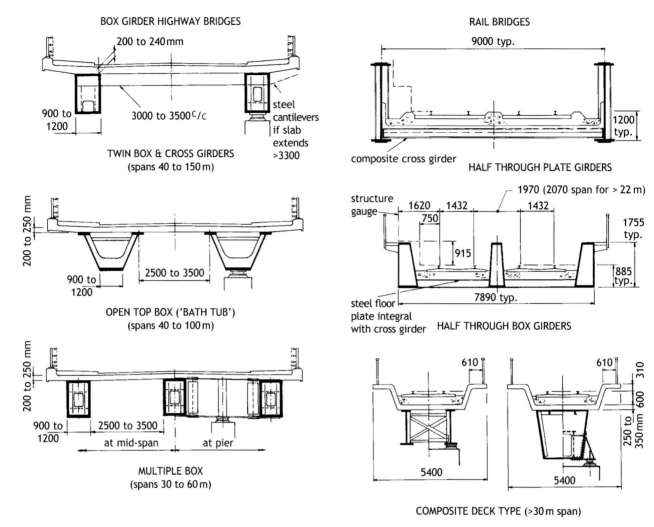

Figure 7.22   Bridges.

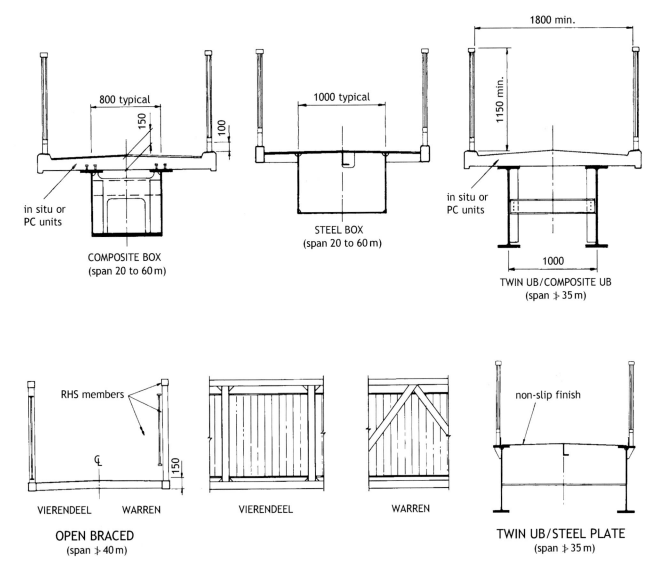

Figure 7.23 Bridges.

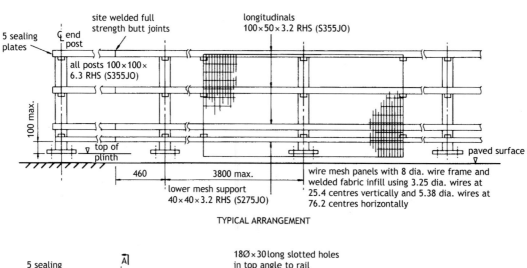

TYPICAL ARRANGEMENT

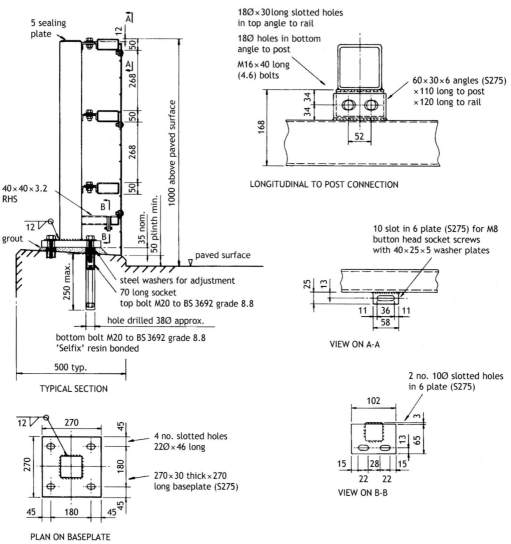

Figure 7.24   Bridges.

note: parapets using this joint should be galvanized

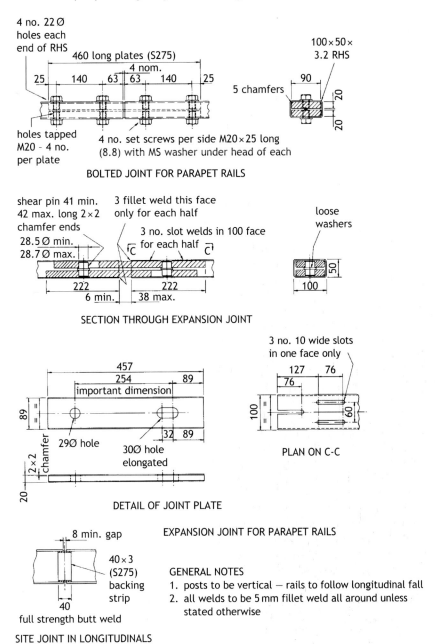

4 no. 22 Ø holes each end of RHS

460 long plates (S275)

4 nom.

25   140   63   63   140   25

holes tapped M20 – 4 no. per plate

4 no. set screws per side M20×25 long (8.8) with MS washer under head of each

100×50×3.2 RHS

90

5 chamfers

20   20

BOLTED JOINT FOR PARAPET RAILS

shear pin 41 min. 42 max. long 2×2 chamfer ends
28.5 Ø min.
28.7 Ø max.

3 fillet weld this face only for each half

3 no. slot welds in 100 face for each half

C   C

222   222
6 min.   38 max.

loose washers

50

100

SECTION THROUGH EXPANSION JOINT

457
254   89
important dimension

89

29 Ø hole

30 Ø hole elongated

32   89

2×2 chamfer

20

DETAIL OF JOINT PLATE

3 no. 10 wide slots in one face only

127   76
76

100

60

PLAN ON C-C

EXPANSION JOINT FOR PARAPET RAILS

8 min. gap

40×3 (S275) backing strip

40

full strength butt weld

SITE JOINT IN LONGITUDINALS

GENERAL NOTES
1. posts to be vertical – rails to follow longitudinal fall
2. all welds to be 5 mm fillet weld all around unless stated otherwise

Figure 7.24 *Contd*

7.8 Single-span highway bridge

The bridge (figures 7.25–7.26 and 7.27) carries a motorway across railway tracks with a clear span of 31.5 m between r.c. abutments and an overall width of 35.02 m. It is suitable for dual three-lane carriageways, hard shoulders and central reserve. It can be adapted to suit different highway widths. Plan curvature of the motorway is accommodated by an increased deck width. Use of steel plate girders with permanent slab formwork allows rapid construction over the railway and would also be suitable across a river. Weathering steel is used to avoid future maintenance painting.

Composite plate girders at 3.08 m centres support the 255 mm thick deck slab and finishes. The edge girders are 1.6 m deep and carry the extra weight of the parapets, which are solid reinforced concrete 'high containment' type. In other locations a lighter open steel parapet is more usual, as shown in figure 7.24.

Inner girders are 1.3 m deep. They are shown fabricated in a single length, but in the UK special permission is required for movement of loads exceeding 27.4 m and this is normally only feasible if good road access is available from the fabrication works, or if rail transport is used. Alternative bolted or welded site splices are shown in figures 7.28 and 7.29. The minimum number of flange thickness changes are made, consistent with available plate lengths. This avoids the high costs of making full penetration butt welds. The girders are precambered in elevation so as to counteract dead load deflection and to follow the road geometry. For calculation of the deflection, girder self weight and concrete slab are assumed carried by the girder alone, whilst finishes and parapets are taken by the composite section. It may be noted that a typical precamber for composite girders is about 0.25–0.5% of span.

Girders are fixed against longitudinal movement at one abutment and free to move at the other. Bearings are proprietary 'pot' or 'disc' type bearings comprising a rubber disc contained within a steel cylinder and piston arrangement. The rubber, being contained, is able to withstand high vertical loads whilst permitting rotation. The free abutment bearings incorporate ptfe (polytetra-fluoroethylene) stainless steel sliding surfaces to cater for thermal movements and concrete shrinkage. Composite steel channel trimmers occur at each abutment to restrain the girders during construction and to stiffen the slab ends. Within the span two lines of transverse channel bracings are provided for erection stability. All site connections are made up using HSFG bolts. For erection the girders were placed in groups of up to three using a lifting beam as shown in figure 7.30. This is convenient where the erection period is limited by short railway occupations and was used to erect the prototype of the bridge described.

The introduction of integral bridges, where the ends of the deck structure and each abutment are continuously joined, has resulted in the omission of deck expansion joints at the abutments, thus minimising the potential consequences of salt chloride attack on the deck slab and substructure.

Drawing Notes

(1) All steel to be weather resistant unpainted to EN 10155 grade S355 J2G1W UON.

(2) All bolts to be HSFG to BS 4395 Part 1. Chemical composition to ASTM A325 Type 3, Grade A, or equivalent weather resistant. To be M24 diameter UON.

(3) Intermediate stiffeners may be radial to camber.

(4) All welds to be fillet welds size 6 mm UON continuous on both sides of all joints.

(5) Butt welds – all transverse welds to flanges and webs to be full penetration welds.

(6) All welding electrodes shall be to BS EN 499. Welds shall possess similar weather-resisting properties to the steel such that these are retained, including possible loss of thickness due to slow rusting. The design allows for loss of thickness of 2 mm on all exposed surfaces.

(7) Temporary lifting cleats may remain in position within slab.

(8) Temporary welds shall not occur within 25 mm of any flange edge.

(9) Complete trial erection of three adjacent plate girders shall be performed. During the trial erection the true relative levels of the steelwork shall be modelled.

(10) The exposed outer surfaces of web top flange and bottom flange, including soffit, to girders 1 and 12, together with all HSFG interfaces, shall be blast cleaned to third quality BS 7079. All other surfaces shall be maintained free from contamination by concrete, mortar, asphalt, paint, oil, grease and any other undesirable contaminants.

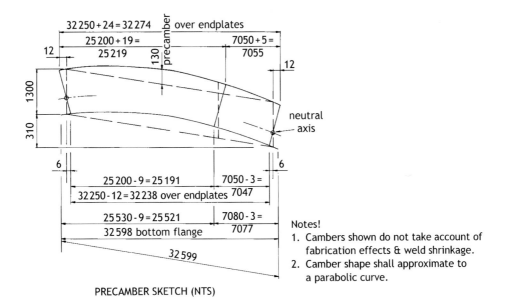

Notes!
1. Cambers shown do not take account of fabrication effects & weld shrinkage.
2. Camber shape shall approximate to a parabolic curve.

PRECAMBER SKETCH (NTS)

Figure 7.25 Single-span highway bridge.

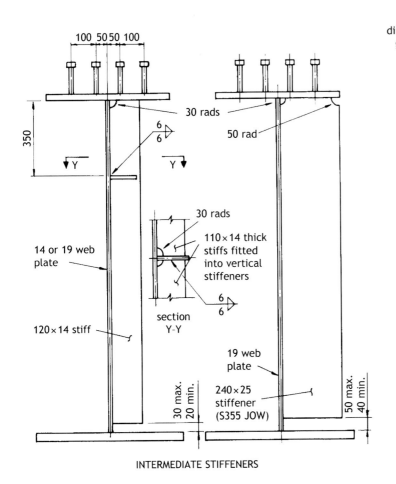

INTERMEDIATE STIFFENERS

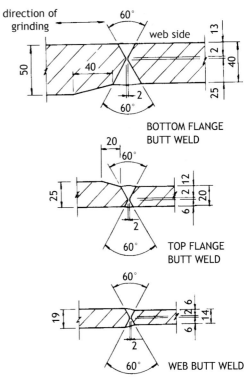

Top surfaces of all bottom flange butt welds to be ground flush

BOTTOM FLANGE BUTT WELD

TOP FLANGE BUTT WELD

WEB BUTT WELD

Precamber at mid-span Girders 2 to 11	
Girder weight	21
Slab etc.	62
Finishes	15
Shrinkage	15
Final precamber	15
Total	128
Specified precamber	130

Figure 7.26 Single-span highway bridge.

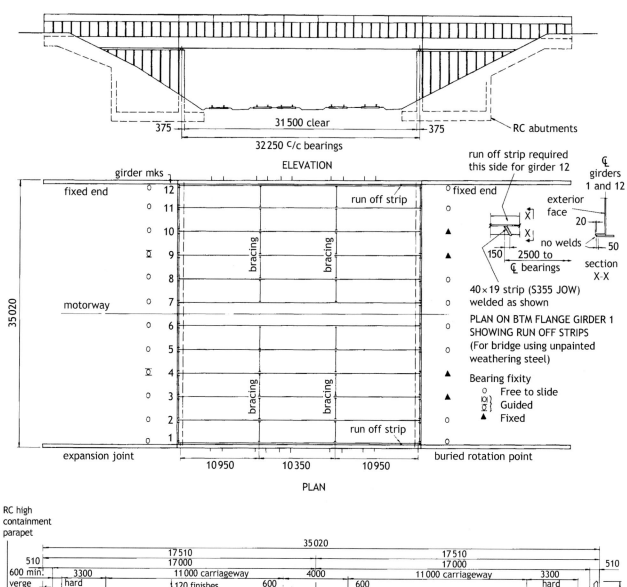

Figure 7.27 Single-span highway bridge.

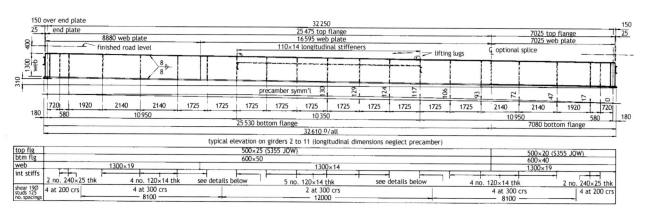

Figure 7.28 Single-span highway bridge.

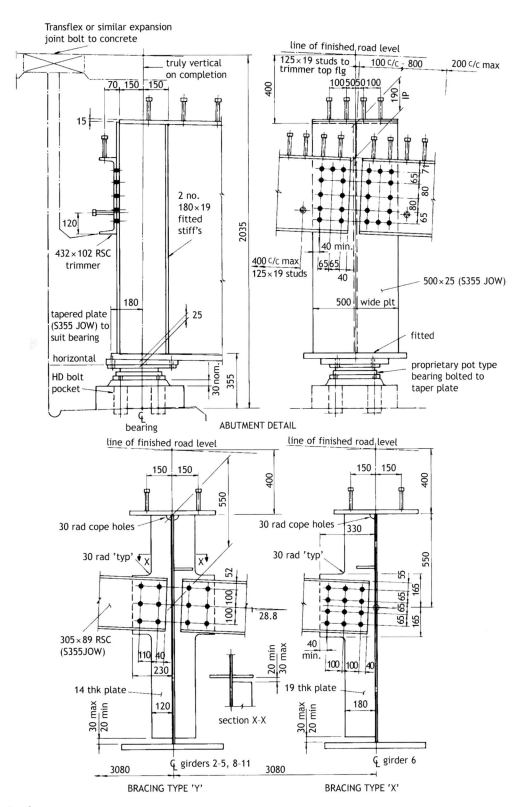

Figure 7.28 *Contd*

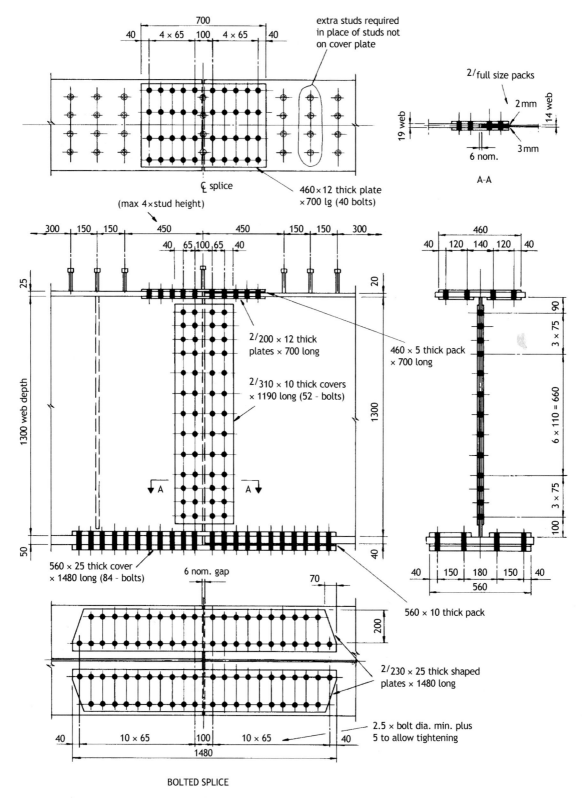

700

40   4 × 65   100   4 × 65   40

extra studs required
in place of studs not
on cover plate

℄ splice

(max 4 × stud height)

460 × 12 thick plate
× 700 lg (40 bolts)

2/full size packs

2 mm

19 web   14 web

3 mm

6 nom.

A-A

300   150   150   450   450   150   150   300

460

40   120   140   120   40

40   65 100 65   40

25   20

90

3 × 75

2/200 × 12 thick
plates × 700 long

460 × 5 thick pack
× 700 long

1300 web depth

2/310 × 10 thick covers
× 1190 long (52 – bolts)

1300

6 × 110 = 660

A   A

3 × 75

50   40

100

560 × 25 thick cover
× 1480 long (84 – bolts)

6 nom. gap

70

40   150   180   150   40

560

560 × 10 thick pack

200

2/230 × 25 thick shaped
plates × 1480 long

40   10 × 65   100   10 × 65   40

2.5 × bolt dia. min. plus
5 to allow tightening

1480

BOLTED SPLICE

Figure 7.29   Single-span highway bridge.

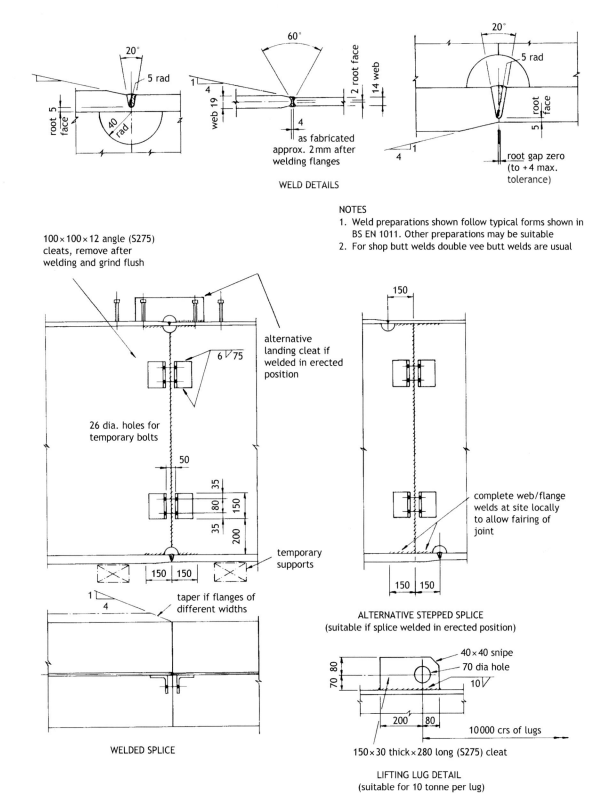

WELD DETAILS

NOTES
1. Weld preparations shown follow typical forms shown in BS EN 1011. Other preparations may be suitable
2. For shop butt welds double vee butt welds are usual

100 × 100 × 12 angle (S275) cleats, remove after welding and grind flush

alternative landing cleat if welded in erected position

26 dia. holes for temporary bolts

temporary supports

taper if flanges of different widths

WELDED SPLICE

complete web/flange welds at site locally to allow fairing of joint

ALTERNATIVE STEPPED SPLICE
(suitable if splice welded in erected position)

40 × 40 snipe
70 dia hole
10000 crs of lugs
150 × 30 thick × 280 long (S275) cleat

LIFTING LUG DETAIL
(suitable for 10 tonne per lug)

Figure 7.29 *Contd*

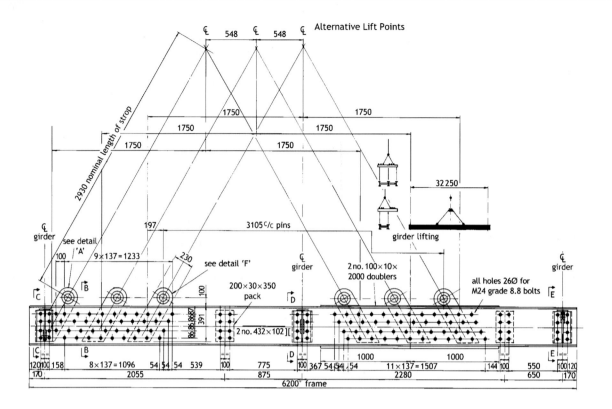

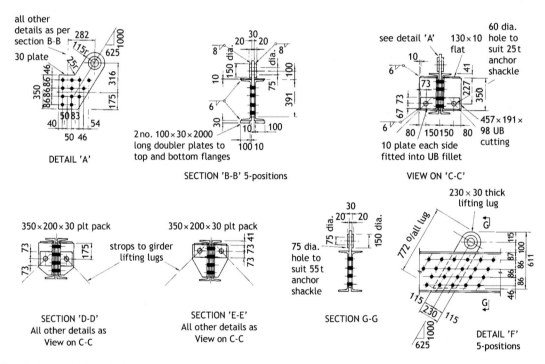

Figure 7.30 Single-span highway bridge.

7.9 Highway sign gantry

In recent years there have been some significant changes to the appearance and structural strength of highway sign gantries. The key differences have been:

- a gradual absence of fixed maintenance access walkways, which have largely contributed to more slender designs,
- a fundamentally different approach to the consequences of vehicle impact, and
- the use of retro-reflective micro-prismatic sheeting for the signs, as an alternative to direct lighting.

Newer gantries can be designed in single or twin span arrangement. The exploitation of the 3-dimensional strength of using a truss girder in the structural configuration can result in significant weight reductions. The resulting lightweight structures can then provide lower fabrication and erection costs. The adaptable arrangement of the front face will allow the fitting of many types and layout of equipment.

The gantry shown in figures 7.31, 7.32 and 7.33 displays advanced direction signs and advanced motorway indicator signs above the three-lane carriageway of a motorway. For larger directional signs on motorways the use of external illumination is possible, with lighting units mounted on a walkway located in front of and below the signs. Such a walkway could also be used for maintenance access, and a heavier type of gantry results.

Square hollow sections (SHS) are used throughout to give a clean appearance. The vertical legs support a square truss girder consisting of main boom chord members and lacings made from SHS members. Welded joints are used throughout except for the leg to end girder connections, which are site bolted using HSFG bolts to ensure the rigid portal action of the gantry.

The typical leg member holding down bolt arrangement is designed to allow rapid erection during a night road closure. This is achieved using a 'bolt box' arrangement located within the concrete base slab. 'Finger' packs can be used so that accurate levelling and securing of the gantry can be achieved, with final grouting of the bases later.

The direction and indicator signs are either mounted internally within the main boom members or externally fitted to vertical support frames, which are mounted above the top chord members of the gantry.

Drawing notes

(1) All steel to be to EN 10210 grade S355 UON.
Hollow sections to be grade S355J2.
(2) Protective treatment.
Grit blast 1st quality after fabrication.
Metal coating – aluminium spray
Paint coats: 1st aluminium epoxy sealer
2nd zinc phosphate CR/alkyd undercoat
3rd zinc phosphate CR/alkyd undercoat
4th MIO CR undercoat
5th CR finish.
Minimum total dry film thickness 250 microns.

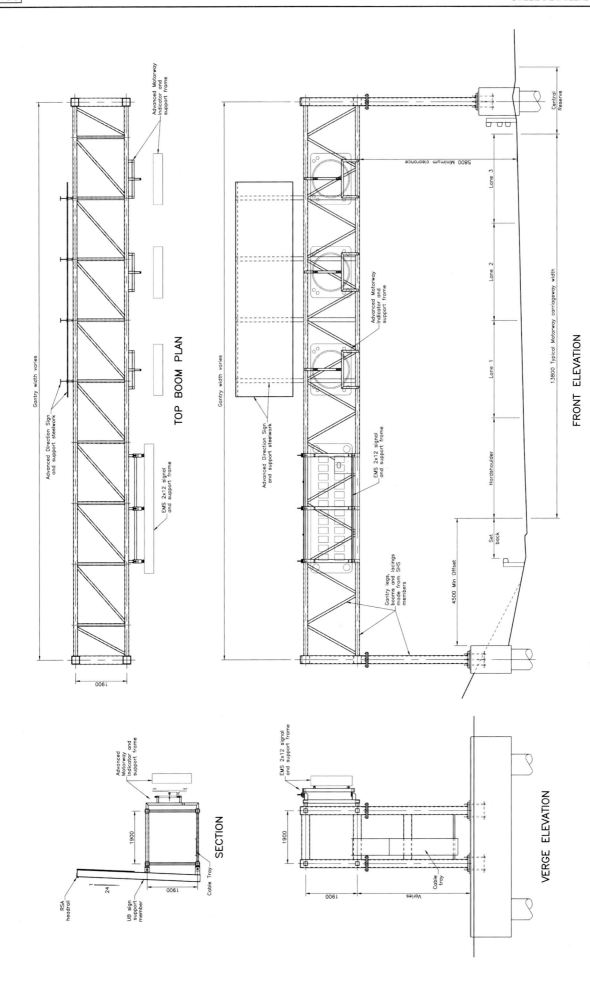

Figure 7.31 Highway sign gantry.

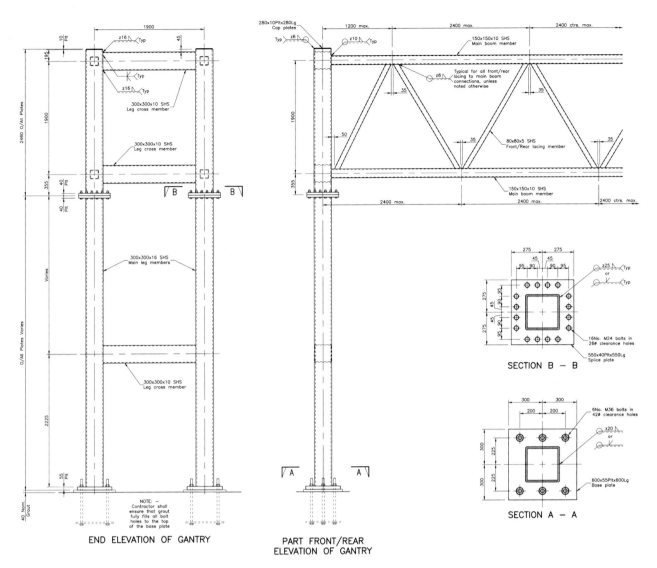

Figure 7.32 Highway sign gantry.

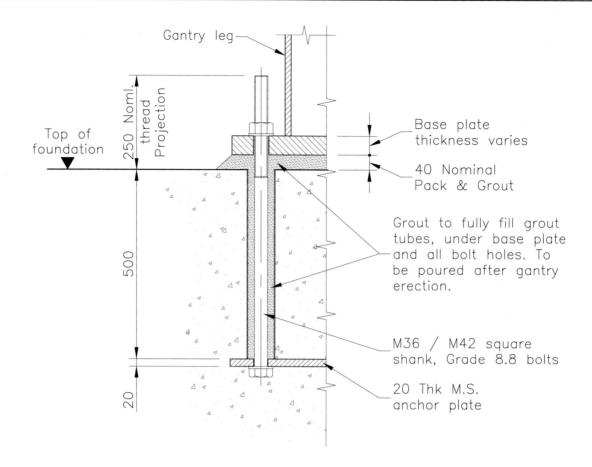

Gantry leg

250 Noml. thread Projection

Top of foundation

Base plate thickness varies

40 Nominal Pack & Grout

Grout to fully fill grout tubes, under base plate and all bolt holes. To be poured after gantry erection.

500

M36 / M42 square shank, Grade 8.8 bolts

20 Thk M.S. anchor plate

20

TYPICAL HOLDING DOWN BOLT DETAILS

Figure 7.33 Highway sign gantry.

7.10 Staircase

The staircase occurs within an industrial complex and is an essential structure. It is typical of many staircases built within factories and would be suitable as fire escape stairs in public buildings. Figure 7.34 shows one landing/flight unit which is connected to similar elements to form a zigzag staircase. Certain design standards relate to staircases regarding proportions of rise (going, length of landings, number of risers between landings, etc.) and these are shown in figure 4.1.

The staircase (figure 7.35) comprises twin steel flat stringers to which are bolted stair treads and tubular handrails. The stringers rely upon the treads to maintain stability against buckling. Channel stringers are also often used. Stair treads and floor/landing panels are of proprietary open bar grating type formed from a series of parallel flat load bearing bars stood on end and equi-spaced with either indented round or square bars. These are resistance welded into the top surface of the load bearing bars primarily to keep them upright. Panels typically 1 m wide and 6 m long or more are supported for elevated walkways and platforms. Normal treatment is galvanizing, which ensures that all interstices receive treatment, but between dip treatment can be used for less corrosive conditions. Stair treads are of similar construction. A number of manufacturers supply this type of flooring.

Handrail standards are proprietary solid forged type with tubular rails made from steel tube to BS EN 1775 grade 13. These are available from several manufacturers for either light or heavy duty applications.

GENERAL NOTES
1. All stair stringer joints to be full strength butt welds.
2. All other welds to be 6 fillet continuous UON.
3. All holes for handrail standards and stair treads to be 14 dia. for M12 grade 4.6 bolts.
4. All other holes are to be 22 dia. for M20 grade 4.6 bolts.
5. All dimensions & details shown for handrailing, standards & treads are to Redman Fisher standard pattern.
6. Standards: 32 dia. solid steel bars.
7. Handrailing: 25 dia. nom. bore tubes. (Spigot joints to be arranged as req'd.)
8. Stairtreads: Ref. FD509 serrated load-bearing bars — 30 × 3 41 pitch × 525 long. Dim.'A' = 249 & Dim.'B' = 100.
9. Finish — all materials to be galvanized except for stair & landing stringers, which are painted as per specification.
10. Materials — to be to EN 10025 (S275) UON.

Figure 7.34 Staircase.

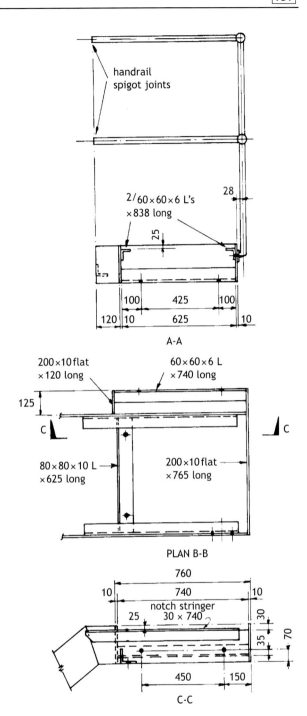

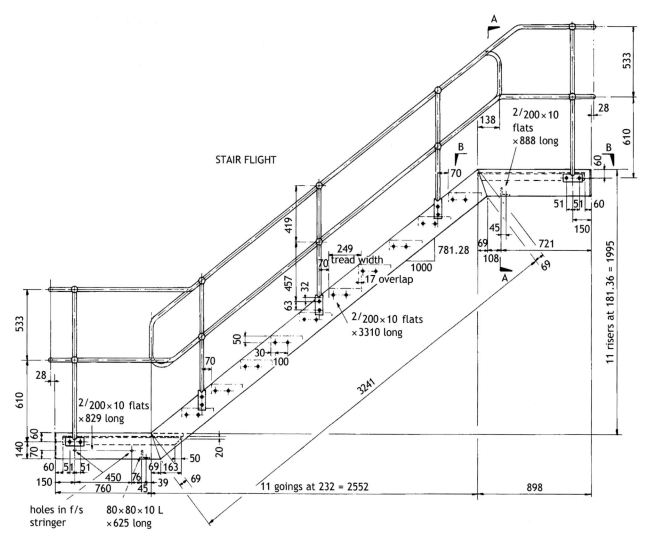

STAIR FLIGHT

2/200×10
flats
×888 long

2/200×10 flats
×3310 long

2/200×10 flats
×829 long

holes in f/s
stringer

80×80×10 L
×625 long

Figure 7.34 *Contd*

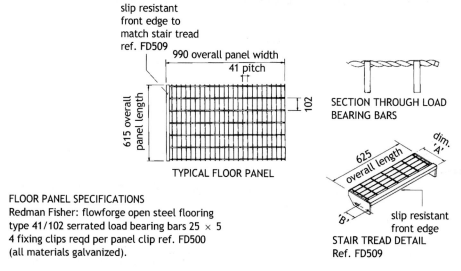

slip resistant
front edge to
match stair tread
ref. FD509

990 overall panel width

41 pitch

615 overall panel length

102

TYPICAL FLOOR PANEL

SECTION THROUGH LOAD
BEARING BARS

625 overall length

dim. 'A'

'B'

slip resistant
front edge

STAIR TREAD DETAIL
Ref. FD509

FLOOR PANEL SPECIFICATIONS
Redman Fisher: flowforge open steel flooring
type 41/102 serrated load bearing bars 25 × 5
4 fixing clips reqd per panel clip ref. FD500
(all materials galvanized).

Figure 7.35 Staircase.

Table of Standards

Codes and Standards referred to in this Edition

BS 4: Part 1: 2005 *Specification for hot rolled sections.*

BS 3692: 2001 *ISO metric precision hexagon bolts, screws and nuts. Specification.*

BS 4190: 2001 *ISO metric black hexagon bolts, screws and nuts. Specification.*

BS 4320: 1968 *Metal washers for general engineering purposes. Metric series.*

BS 4395 *High strength friction grip bolts and associated nuts and washers for structural engineering.*

Part 1: 1969 General grade.

Part 2: 1969 Higher grade.

BS 4604 *The use of high strength friction grip bolts in structural steelwork.*

Part 1: 1970 General grade.

Part 2: 1970 Higher grade.

BS 5400 *Steel, concrete and composite bridges.*

Part 1: 1988 General statement.

Part 2: 2006 Specification for loads.

Part 3: 2000 Code of practice for design of steel bridges.

Part 5: 2005 Code of practice for design of composite bridges.

Part 6: 1999 Specification for materials and workmanship, steel.

Part 9.1: 1983 Bridge bearings. Code of practice for design of bridge bearings. *(This section partially replaced by BS EN 1337–4, and BS EN 1337–6, but remains current).*

Part 9.2: 1983 Bridge bearings. Specification for materials, manufacture and installation of bridge bearings. *(This section partially replaced by BS EN 1337–2, BS EN 1337–3, BS EN 1337–5, and BS EN 1337–7).*

Part 10: 1980 Code of practice for fatigue.

Part 10C: 1999 Charts for classification of details for fatigue.

BS 5950 *Structural use of steelwork in building.* (This standard is now withdrawn by BSI, but can still continue to be used on any existing structural projects).

Part 1: 2000 Code of practice for design – Rolled and welded sections.

Part 2: 2001 Specification for materials, fabrication and erection – Rolled and welded sections.

Part 3: 1990 Design in composite construction. Section 3.1: Code of practice for design of simple and continuous composite beams.

Part 4: 1994 Code of practice for design of composite slabs with profiled steel sheeting.

BS 7079-0: 2009 *General introduction to standards for preparation of steel substrates before application of paints and related products. Introduction.* (Reference should also be made to other parts of BS 7079 and to BS EN ISO 8502, BS EN ISO 8502 and BS EN ISO 11124).

BS 7668: 1994 *Specification for weldable structural steels. Hot finished structural hollow sections in weather resistant steels.*

BS EN 1011 *Welding. Recommendations for welding of metallic materials.*

Part 1: 2009 General guidance for arc welding.

Part 2: 2001 Arc welding of ferritic steels.

BS EN 1090 *Execution of steel structures and aluminium structures.*

Part 2: 2008 Technical requirements for the execution of steel structures.

BS EN 1991: Eurocode 1: *Actions on structures* (each part has a relevant UK National Annex).

Part 1-1: 2002 General actions - Densities, self-weight, imposed loads for buildings.

BS EN 1993: Eurocode 3: *Design of steel structures* (each part has a relevant UK National Annex).

Part 1-1: 2005 General rules and rules for buildings.

Steel Detailers' Manual, Third Edition. Alan Hayward and Frank Weare. Third edition revised by Anthony Oakhill.
© 2011 Alan Hayward, Frank Weare and Anthony Oakhill. Published 2011 by Blackwell Publishing Ltd.

Part 1-2: 2005 General rules – Structural fire design

Part 1-8: 2005 Design of joints

Part 1-10: 2005 Material toughness and through-thickness properties

Part 2: 2006 Bridges

BS EN 1994: Eurocode 4: Design of composite steel and concrete structures (each part has a relevant UK National Annex).

Part 1-1: 2004 General rules and rules for buildings.

Part 2-1: 2005 General rules and rules for bridges.

BS EN 10025: 2004 *Hot rolled products of non-alloy structural steels. Technical delivery conditions.*

BS EN 10210: *Hot finished structural hollow sections of non-alloy and fine grain structural steels.*

Part 1: 2006 Technical delivery requirements.

BS EN 10219: *Cold formed welded structural hollow sections of non-alloy and fine grain steels.*

Part 1: 2006 Technical delivery requirements.

BS EN 14399: *High strength structural bolting assemblies for preloading.*

Part 1: 2005 General requirements

Part 3: 2005 System HR. Hexagon bolt and nut assemblies.

BS EN 22553: 1995 *Welded, brazed and soldered joints. Symbolic representation on drawings.*

BS EN ISO 2560: 2009 *Welding consumables. Covered electrodes for manual metal arc welding of non alloy and fine grain steels. Classification.*

BS EN ISO 4063: 2009 *Welding and allied processes. Nomenclature of processes and reference numbers.*

BS EN ISO 4157: 1999 *Construction drawings. Designation systems.* (Reference should also be made to BS EN ISO 8560 and BS EN ISO 9431).

BS EN ISO 14713: 2009 *Protection against corrosion of iron and steel in structures. Zinc and aluminium coatings. Introduction.*

References

1 Departmental Standard BD7/81. *Weathering steel for highway structures*. Department of Transport, 1981.
2 Swedish Standard SIS 05 59 00. *Rust grades for steel surfaces and preparation grades prior to protective coating*. Swedish Standards Commission, Stockholm, 1971.
3 *Steel structures painting manual*. Steel Structures Painting Council, Pittsburgh, USA.
   Volume 1: 1966 Good painting practice.
   Volume 2: 1973 systems and specification.

4 *Notes for guidance on the specification for highway works. Series NG 1900 Protection of steelwork against corrosion*. Highways Agency, HMSO, 1998 (with later amendments).
5 *Steelwork design guide to BS 5950-1: 2000*.
   Volume 1: Section properties and member capacities, SCI, 2001.
6 *Structural fasteners and their application*, BCSA, 1978.
7 Fire protection for structural steel in buildings, SCI, 1992.

Steel Detailers' Manual, Third Edition. Alan Hayward and Frank Weare. Third edition revised by Anthony Oakhill.
© 2011 Alan Hayward, Frank Weare and Anthony Oakhill. Published 2011 by Blackwell Publishing Ltd.

Further Reading

Design

(1) *Steel Designers' Manual* 6th Edn (revised). SCI and Wiley-Blackwell, 2005.

(2) BS 6399 *Loading for buildings.*
Part 1: 1996. Code of practice for dead and imposed loads.

(3) BS 5502 *Buildings and structures for agriculture.*
Part 21: 1990 Code of practice for selection and use of construction materials.
Part 22: 2003 Code of practice for design, construction and loading.

(4) BS 2573 *Rules for the design of cranes.*
Part 1: 1983 Specification for classification, stress calculations and design criteria for structures.

(5) BS 2853: 1957 *Specification for the design and testing of steel overhead runway beams.*

(6) BS 466: 1984 *Specification for power driven overhead travelling cranes, semi-goliath and goliath cranes for general use.*

(7) 105 *Factory Stairways, Ladders and Handrails (including Access Platforms and Ramps).* 7th Edn. Engineering Equipment and Materials Users Association, 2007.

(8) BS 5395: *Stairs.*
Part 1: 2010 Code of practice for the design of stairs with straight flights and winders.

(9) BS 4592-0: 2006 *Industrial type flooring and stair treads. Common design requirements and recommendations for installation.*

(10) *Joints in Steel Construction Simple Connections,* BCSA and SCI, 2002.

(11) *Joints in Steel Construction: Moment Connections,* SCI and BCSA, 1995.

(12) *Joints in Steel Construction: Composite Connections,* SCI and BCSA, 1998.

Detailing

(1) *Steel Details*, Publication No. 41/05, BCSA, 2005

(2) BS 8888: 2008 *Technical product specification. Specifications.*

(3) BS EN ISO 1660: 1996 *Technical drawings. Dimensioning and tolerancing of profiles.*

(4) BS EN ISO 7083: 1995 *Technical drawings. Symbols for geometrical tolerancing. Proportions and dimensions.*

Steel sections

(1) *Handbook of structural steelwork. Properties and safe load tables.* 4th Edn. BCSA and SCI, 2007.

(2) *Advance sections, CE marked structural sections.* Corus, 2006.

(3) *Steel bearing piles guide*, SCI, 1997.

Protective treatment

(1) BS EN ISO 1461: 2009 *Hot dip galvanized coatings on fabricated iron and steel articles. Specifications and test methods.*

(2) BS EN ISO 2081: 2008 *Metallic and other inorganic coatings. Electroplated coatings of zinc with supplementary treatments on iron or steel.*

(3) BS EN ISO 2082: 2008 *Metallic coatings. Electroplated coatings of cadmium with supplementary treatments on iron or steel.*

(4) BS EN ISO 2063: 2005 *Thermal spraying. Metallic and other inorganic coatings. Zinc, aluminium and their alloys.*

(5) BS 7371: *Coatings on metal fasteners.*
Part 1: 2009 Specification for general requirements and selection guidelines

Steel Detailers' Manual, Third Edition. Alan Hayward and Frank Weare. Third edition revised by Anthony Oakhill.
© 2011 Alan Hayward, Frank Weare and Anthony Oakhill. Published 2011 by Blackwell Publishing Ltd.

(6) *Steel bridges–Material matters, Corrosion protection.* Corus, 2010.

Erection

(1) BS 5975: 2008 *Code of practice for temporary works procedures and the permissible stress design of falsework.*

(2) *Code of Practice for Erection of Low Rise Buildings.* BCSA, 2004.

(3) *Guide to the Erection of Multi-Storey Buildings.* BCSA, 2006

(4) Health and Safety Executive Guidance Notes GS28.

(5) *Steel Buildings.* BCSA Publication No. 35/03.

Composite construction

(1) *Composite structures of steel and concrete*, R.P. Johnson and R.J. Buckby. Volume 1 Beams, slabs, columns, and frames for buildings. 3rd Edn, Wiley-Blackwell, 2004. Volume 2 Bridges. 2nd Edn. Collins (now Wiley-Blackwell), 1986

Bridges

(1) *Steel Bridges.* 3rd Edn. BCSA Publication No. 51/10.

(2) *International symposium on steel bridges*, ECCS/BCSA, Publication No. E97/96, Rotterdam, 1996.

(3) *Composite Steel Highway Bridges*, A.C.G. Hayward, D. C. Iles Corus, Revised 2005.

(4) *The Design of Steel Footbridges*, D.C. Iles. Corus, Revised 2005.

International

(1) *Iron and steel specifications*, 7th Edn, British Steel plc (now Corus), 1989.

Welding

(1) *Introduction to the welding of structural steelwork*, J.L. Pratt, 3rd Edn. SCI, 1989.

(2) ANSI/AWS D1, 1–81 *Structural welding code*, USA.

(3) BS EN 756: 2004 *Welding consumables. Solid wires, solid wire-flux and tubular cored electrode flux combinations for submerged arc welding of non alloy and fine grain steels. Classification.*

(4) BS EN 760: 1996 *Welding consumables. Fluxes for submerged arc welding. Classification.*

(5) BS 499 *Welding terms and symbols.*
Part 1: 2009 Glossary for welding, brazing and thermal cutting.
Part 2: 1999 European arc welding symbols in chart form.

Weld testing

(1) BS EN ISO 15614 *Specification and qualification of welding procedures for metallic materials.*
Part 1: 2004 Welding procedure test. Arc gas welding of steels and arc welding of nickel and nickel alloys.

(2) BS EN 287 *Qualification test of welders.*
Part 1: 2004 Fusion welding. Steels.

(3) BS 4872 *Specification for test of welders when welding procedure approval is not required.*
Part 1: 1982 Fusion welding. Steels.

(4) BS EN 1321: 1997 *Destructive test on welds in metallic materials. Macroscopic and microscopic examination of welds.*

(5) BS EN 1435: 1997 *Non-destructive examination of welds. Radiographic examination of welded joints.*

(6) BS EN 1714: 1998 *Non-destructive testing of welded joints. Ultrasonic testing of welded joints.*

(7) BS EN ISO 9934 *Non-destructive testing.*
Part 1: 2001 Magnetic particle tests. General principles.

(8) BS EN 571. *Non-destructive testing.*
Part 1: 1997 Penetrant testing. General principles.

(9) BS EN 970: 1997 *Non-destructive examination of fusion welds. Visual examination.*

(10) BS 7910: 2005 *Guide to methods for assessing the acceptability of flaws in metallic structures.*

Abbreviations

AISC   American Institute of Steel Construction, One East Wacker Drive, Suite 700, Chicago, IL, 60601–1802.

BCSA   British Constructional Steelwork Association Limited, 4 Whitehall Court, Westminster, London SW1A 2ES.

BS   British Standard – British Standards may be obtained from: British Standards Institution, 389 Chiswick High Road, London W4 4AL.

Tata   Tata Steel Construction Services and Development, PO Box 1, Brigg Road, Scunthorpe, North Lincolnshire DN16 1BP.

SCI   Steel Construction Institute, Silwood Park, Ascot, Berkshire SL5 7QN.

Appendix

The Appendix contains useful information including weights of bars and flats, conversion factors and trigonometrical expressions.

Steel Detailers' Manual, Third Edition. Alan Hayward and Frank Weare. Third edition revised by Anthony Oakhill.

Mass of round and square bars

Kilogrammes per linear metre

Dia. or side	Round	Square	Dia. or side	Round	Square	Dia. or side	Round	Square
mm	●	■	mm	●	■	mm	●	■
10	0.62	0.79	45	12.48	15.90	100	61.65	78.50
11	0.75	0.95	46	13.05	16.61	105	67.97	86.55
12	0.89	1.13	47	13.62	17.34	110	74.60	94.90
13	1.04	1.33	48	14.21	18.09	115	81.54	103.82
14	1.21	1.54	49	14.80	18.85	120	88.78	113.04
15	1.39	1.77	50	15.41	19.63	125	96.33	122.66
16	1.58	2.01	51	16.04	20.42	130	104.19	132.67
17	1.78	2.27	52	16.67	21.23	135	112.36	143.07
18	2.00	2.54	53	17.32	22.05	140	120.84	153.86
19	2.23	2.83	54	17.98	22.89	145	129.63	165.05
20	2.47	3.14	55	18.65	23.75	150	138.72	176.63
21	2.72	3.46	56	19.33	24.62	155	148.12	188.60
22	2.98	3.80	57	20.03	25.50	160	157.83	200.96
23	3.26	4.15	58	20.74	26.41	165	167.85	213.72
24	3.55	4.52	59	21.46	27.33	170	178.18	226.87
25	3.85	4.91	60	22.20	28.26	175	188.81	240.41
26	4.17	5.31	61	22.94	29.21	180	199.76	254.34
27	4.49	5.72	62	23.70	30.18	185	211.01	268.67
28	4.83	6.15	63	24.47	31.16	190	222.57	283.39
29	5.19	6.60	64	25.25	32.15	195	234.44	298.50
30	5.55	7.07	65	26.05	33.17	200	246.62	314.00
31	5.92	7.54	66	26.86	34.19	205	259.10	329.90
32	6.31	8.04	67	27.68	35.24	210	271.89	346.19
33	6.71	8.55	68	28.51	36.30	215	284.99	362.87
34	7.13	9.07	69	29.35	37.37	220	298.40	379.94
35	7.55	9.62	70	30.21	38.47	225	312.12	397.41
36	7.99	10.17	71	31.08	39.57	230	326.15	415.27
37	8.44	10.75	72	31.96	40.69	235	340.48	433.52
38	8.90	11.34	73	32.86	41.83	240	355.13	452.16
39	9.38	11.94	74	33.76	42.99	250	385.34	490.63
40	9.86	12.56	75	34.68	44.16	260	416.78	530.66
41	10.36	13.20	80	39.46	50.24	270	449.46	572.27
42	10.88	13.85	85	44.54	56.72	280	483.37	615.44
43	11.40	14.51	90	49.94	63.59	290	518.51	660.19
44	11.94	15.20	95	55.64	70.85	300	554.88	706.50

Suppliers should be consulted regarding availability of sizes.

Mass of flats

Kilogrammes per linear metre

Width mm	Thickness in millimetres															
	1	2	3	4	5	6	7	8	9	10	15	20	25	30	40	50
5	0.04	0.08	0.12	0.16	0.20	0.24	0.27	0.31	0.35	0.39	0.59	0.79	0.98	1.18	1.57	1.96
10	0.08	0.16	0.24	0.31	0.39	0.47	0.55	0.63	0.71	0.79	1.18	1.57	1.96	2.36	3.14	3.93
15	0.12	0.24	0.35	0.47	0.59	0.71	0.82	0.94	1.06	1.18	1.77	2.36	2.94	3.53	4.71	5.89
20	0.16	0.31	0.47	0.63	0.79	0.94	1.10	1.26	1.41	1.57	2.36	3.14	3.93	4.71	6.28	7.85
25	0.20	0.39	0.59	0.79	0.98	1.18	1.37	1.57	1.77	1.96	2.94	3.93	4.91	5.89	7.85	9.81
30	0.24	0.47	0.71	0.94	1.18	1.41	1.65	1.88	2.12	2.36	3.53	4.71	5.89	7.07	9.42	11.8
35	0.27	0.55	0.82	1.10	1.37	1.65	1.92	2.20	2.47	2.75	4.12	5.50	6.87	8.24	11.0	13.7
40	0.31	0.63	0.94	1.26	1.57	1.88	2.20	2.51	2.83	3.14	4.71	6.28	7.85	9.42	12.6	15.7
45	0.35	0.71	1.06	1.41	1.77	2.12	2.47	2.83	3.18	3.53	5.30	7.07	8.83	10.6	14.1	17.7
50	0.39	0.79	1.18	1.57	1.96	2.36	2.75	3.14	3.53	3.93	5.89	7.85	9.81	11.8	15.7	19.6
55	0.43	0.86	1.30	1.73	2.16	2.59	3.02	3.45	3.89	4.32	6.48	8.64	10.8	13.0	17.3	21.6
60	0.47	0.94	1.41	1.88	2.36	2.83	3.30	3.77	4.24	4.71	7.07	9.42	11.8	14.1	18.8	23.6
65	0.51	1.02	1.53	2.04	2.55	3.06	3.57	4.08	4.59	5.10	7.65	10.2	12.8	15.3	20.4	25.5
70	0.55	1.10	1.65	2.20	2.75	3.30	3.85	4.40	4.95	5.50	8.24	11.0	13.7	16.5	22.0	27.5
75	0.59	1.18	1.77	2.36	2.94	3.53	4.12	4.71	5.30	5.89	8.83	11.8	14.7	17.7	23.6	29.4
80	0.63	1.26	1.88	2.51	3.14	3.77	4.40	5.02	5.65	6.28	9.42	12.6	15.7	18.8	25.1	31.4
85	0.67	1.33	2.00	2.67	3.34	4.00	4.67	5.34	6.01	6.67	10.0	13.3	16.7	20.0	26.7	33.4
90	0.71	1.41	2.12	2.83	3.53	4.24	4.95	5.65	6.36	7.07	10.6	14.1	17.7	21.2	28.3	35.3
95	0.75	1.49	2.24	2.98	3.73	4.47	5.22	5.97	6.71	7.46	11.2	14.9	18.6	22.4	29.8	37.3
100	0.79	1.57	2.36	3.14	3.93	4.71	5.50	6.28	7.07	7.85	11.8	15.7	19.6	23.6	31.4	39.3
110	0.86	1.73	2.59	3.45	4.32	5.18	6.04	6.91	7.77	8.64	13.0	17.3	21.6	25.9	34.5	43.2
120	0.94	1.88	2.83	3.77	4.71	5.65	6.59	7.54	8.48	9.42	14.1	18.8	23.6	28.3	37.7	47.1
130	1.02	2.04	3.06	4.08	5.10	6.12	7.14	8.16	9.18	10.2	15.3	20.4	25.5	30.6	40.8	51.0
140	1.10	2.20	3.30	4.40	5.50	6.59	7.69	8.79	9.89	11.0	16.5	22.0	27.5	33.0	44.0	55.0
150	1.18	2.36	3.53	4.71	5.89	7.07	8.24	9.42	10.6	11.8	17.7	23.6	29.4	35.3	47.1	58.9
160	1.26	2.51	3.77	5.02	6.28	7.54	8.79	10.0	11.3	12.6	18.8	25.1	31.4	37.7	50.2	62.8
170	1.33	2.67	4.00	5.34	6.67	8.01	9.34	10.7	12.0	13.3	20.0	26.7	33.4	40.0	53.4	66.7
180	1.41	2.83	4.24	5.65	7.07	8.48	9.89	11.3	12.7	14.1	21.2	28.3	35.3	42.4	56.5	70.7
190	1.49	2.98	4.47	5.97	7.46	8.95	10.4	11.9	13.4	14.9	22.4	29.8	37.3	44.7	59.7	74.6
200	1.57	3.14	4.71	6.28	7.85	9.42	11.0	12.6	14.1	15.7	23.6	31.4	39.3	47.1	62.8	78.5
210	1.65	3.30	4.95	6.59	8.24	9.89	11.5	13.2	14.8	16.5	24.7	33.0	41.2	49.5	65.9	82.4
220	1.73	3.45	5.18	6.91	8.64	10.4	12.1	13.8	15.5	17.3	25.9	34.5	43.2	51.8	69.1	86.4
230	1.81	3.61	5.42	7.22	9.03	10.8	12.6	14.4	16.2	18.1	27.1	36.1	45.1	54.2	72.2	90.3
240	1.88	3.77	5.65	7.54	9.42	11.3	13.2	15.1	17.0	18.8	28.3	37.7	47.1	56.5	75.4	94.2
250	1.96	3.93	5.89	7.85	9.81	11.8	13.7	15.7	17.7	19.6	29.4	39.3	49.1	58.9	78.5	98.1

For actual widths and thicknesses available, application should be made to manufacturers. Masses for greater widths and/or thicknesses than those tabulated may be obtained by appropriate addition from the range of masses given.

Mass of flats *Contd*

Kilogrammes per linear metre

Width mm	\multicolumn{18}{c}{Thickness in millimetres}

Width mm	1	2	3	4	5	6	7	8	9	10	15	20	25	30	40	50
260	2.04	4.08	6.12	8.16	10.2	12.2	14.3	16.3	18.4	20.4	30.6	40.8	51.0	61.2	81.6	102
270	2.12	4.24	6.36	8.48	10.6	12.7	14.8	17.0	19.1	21.2	31.8	42.4	53.0	63.6	84.8	106
280	2.20	4.40	6.59	8.79	11.0	13.2	15.4	17.6	19.8	22.0	33.0	44.0	55.0	65.9	87.9	110
290	2.28	4.55	6.83	9.11	11.4	13.7	15.9	18.2	20.5	22.8	34.1	45.5	56.9	68.3	91.1	114
300	2.36	4.71	7.07	9.42	11.8	14.1	16.5	18.8	21.2	23.6	35.3	47.1	58.9	70.7	94.2	118
310	2.43	4.87	7.30	9.73	12.2	14.6	17.0	19.5	21.9	24.3	36.5	48.7	60.8	73.0	97.3	122
320	2.51	5.02	7.54	10.0	12.6	15.1	17.6	20.1	22.6	25.1	37.7	50.2	62.8	75.4	100	126
330	2.59	5.18	7.77	10.4	13.0	15.5	18.1	20.7	23.3	25.9	38.9	51.8	64.8	77.7	104	130
340	2.67	5.34	8.01	10.7	13.3	16.0	18.7	21.4	24.0	26.7	40.0	53.4	66.7	80.1	107	133
350	2.75	5.50	8.24	11.0	13.7	16.5	19.2	22.0	24.7	27.5	41.2	55.0	68.7	82.4	110	137
360	2.83	5.65	8.48	11.3	14.1	17.0	19.8	22.6	25.4	28.3	42.4	56.5	70.7	84.8	113	141
370	2.90	5.81	8.71	11.6	14.5	17.4	20.3	23.2	26.1	29.0	43.6	58.1	72.6	87.1	116	145
380	2.98	5.97	8.95	11.9	14.9	17.9	20.9	23.9	26.8	29.8	44.7	59.7	74.6	89.5	119	149
390	3.06	6.12	9.18	12.2	15.3	18.4	21.4	24.5	27.6	30.6	45.9	61.2	76.5	91.8	122	153
400	3.14	6.28	9.42	12.6	15.7	18.8	22.0	25.1	28.3	31.4	47.1	62.8	78.5	94.2	126	157
410	3.22	6.44	9.66	12.9	16.1	19.3	22.5	25.7	29.0	32.2	48.3	64.4	80.5	96.6	129	161
420	3.30	6.59	9.89	13.2	16.5	19.8	23.1	26.4	29.7	33.0	49.5	65.9	82.4	98.9	132	165
430	3.38	6.75	10.1	13.5	16.9	20.3	23.6	27.0	30.4	33.8	50.6	67.5	84.4	101	135	169
440	3.45	6.91	10.4	13.8	17.3	20.7	24.2	27.6	31.1	34.5	51.8	69.1	86.4	104	138	173
450	3.53	7.07	10.6	14.1	17.7	21.2	24.7	28.3	31.8	35.3	53.0	70.7	88.3	106	141	177
460	3.61	7.22	10.8	14.4	18.1	21.7	25.3	28.9	32.5	36.1	54.2	72.2	90.3	108	144	181
470	3.69	7.38	11.1	14.8	18.4	22.1	25.8	29.5	33.2	36.9	55.3	73.8	92.2	111	148	184
480	3.77	7.54	11.3	15.1	18.8	22.6	26.4	30.1	33.9	37.7	56.5	75.4	94.2	113	151	188
490	3.85	7.69	11.5	15.4	19.2	23.1	26.9	30.8	34.6	38.5	57.7	76.9	96.2	115	154	192
500	3.93	7.85	11.8	15.7	19.6	23.6	27.5	31.4	35.3	39.3	58.9	78.5	98.1	118	157	196
510	4.00	8.01	12.0	16.0	20.0	24.0	28.0	32.0	36.0	40.0	60.1	80.1	100	120	160	200
520	4.08	8.16	12.2	16.3	20.4	24.5	28.6	32.7	36.7	40.8	61.2	81.6	102	122	163	204
530	4.16	8.32	12.5	16.6	20.8	25.0	29.1	33.3	37.4	41.6	62.4	83.2	104	125	166	208
540	4.24	8.48	12.7	17.0	21.2	25.4	29.7	33.9	38.2	42.4	63.6	84.8	106	127	170	212
550	4.32	8.64	13.0	17.3	21.6	25.9	30.2	34.5	38.9	43.2	64.8	86.4	108	130	173	216
560	4.40	8.79	13.2	17.6	22.0	26.4	30.8	35.2	39.6	44.0	65.9	87.9	110	132	176	220
570	4.47	8.95	13.4	17.9	22.4	26.8	31.3	35.8	40.3	44.7	67.1	89.5	112	134	179	224
580	4.55	9.11	13.7	18.2	22.8	27.3	31.9	36.4	41.0	45.5	68.3	91.1	114	137	182	228
590	4.63	9.26	13.9	18.5	23.2	27.8	·32.4	37.1	41.7	46.3	69.5	92.6	116	139	185	232
600	4.71	9.42	14.1	18.8	23.6	28.3	33.0	37.7	42.4	47.1	70.7	94.2	118	141	188	236

Mass of flats *Contd*

Kilogrammes per linear metre

Width	Thickness in millimetres															
mm	1	2	3	4	5	6	7	8	9	10	15	20	25	30	40	50
610	4.79	9.58	14.4	19.2	23.9	28.7	33.5	38.3	43.1	47.9	71.8	95.8	120	144	192	239
620	4.87	9.73	14.6	19.5	24.3	29.2	34.1	38.9	43.8	48.7	73.0	97.3	122	146	195	243
630	4.95	9.89	14.8	19.8	24.7	29.7	34.6	39.6	44.5	49.5	74.2	98.9	124	148	198	247
640	5.02	10.0	15.1	20.1	25.1	30.1	35.2	40.2	45.2	50.2	75.4	100	126	151	201	251
650	5.10	10.2	15.3	20.4	25.5	30.6	35.7	40.8	45.9	51.0	76.5	102	128	153	204	255
660	5.18	10.4	15.5	20.7	25.9	31.1	36.3	41.4	46.6	51.8	77.7	104	130	155	207	259
670	5.26	10.5	15.8	21.0	26.3	31.6	36.8	42.1	47.3	52.6	78.9	105	131	158	210	263
680	5.34	10.7	16.0	21.4	26.7	32.0	37.4	42.7	48.0	53.4	80.1	107	133	160	214	267
690	5.42	10.8	16.2	21.7	27.1	32.5	37.9	43.3	48.7	54.2	81.2	108	135	162	217	271
700	5.50	11.0	16.5	22.0	27.5	33.0	38.5	44.0	49.5	55.0	82.4	110	137	165	220	275
710	5.57	11.1	16.7	22.3	27.9	33.4	39.0	44.6	50.2	55.7	83.6	111	139	167	223	279
720	5.65	11.3	17.0	22.6	28.3	33.9	39.6	45.2	50.9	56.5	84.8	113	141	170	226	283
730	5.73	11.5	17.2	22.9	28.7	34.4	40.1	45.8	51.6	57.3	86.0	115	143	172	229	287
740	5.81	11.6	17.4	23.2	29.0	34.9	40.7	46.5	52.3	58.1	87.1	116	145	174	232	290
750	5.89	11.8	17.7	23.6	29.4	35.3	41.2	47.1	53.0	58.9	88.3	118	147	177	236	294
760	5.97	11.9	17.9	23.9	29.8	35.8	41.8	47.7	53.7	59.7	89.5	119	149	179	239	298
770	6.04	12.1	18.1	24.2	30.2	36.3	42.3	48.4	54.4	60.4	90.7	121	151	181	242	302
780	6.12	12.2	18.4	24.5	30.6	36.7	42.9	49.0	55.1	61.2	91.8	122	153	184	245	306
790	6.20	12.4	18.6	24.8	31.0	37.2	43.4	49.6	55.8	62.0	93.0	124	155	186	248	310
800	6.28	12.6	18.8	25.1	31.4	37.7	44.0	50.2	56.5	62.8	94.2	126	157	188	251	314
810	6.36	12.7	19.1	25.4	31.8	38.2	44.5	50.9	57.2	63.6	95.4	127	159	191	254	318
820	6.44	12.9	19.3	25.7	32.2	38.6	45.1	51.5	57.9	64.4	96.6	129	161	193	257	322
830	6.52	13.0	19.5	26.1	32.6	39.1	45.6	52.1	58.6	65.2	97.7	130	163	195	261	326
840	6.59	13.2	19.8	26.4	33.0	39.6	46.2	52.8	59.3	65.9	98.9	132	165	198	264	330
850	6.67	13.3	20.0	26.7	33.4	40.0	46.7	53.4	60.1	66.7	100	133	167	200	267	334
860	6.75	13.5	20.3	27.0	33.8	40.5	47.3	54.0	60.8	67.5	101	135	169	203	270	338
870	6.83	13.7	20.5	27.3	34.1	41.0	47.8	54.6	61.5	68.3	102	137	171	205	273	341
880	6.91	13.8	20.7	27.6	34.5	41.4	48.4	55.3	62.2	69.1	104	138	173	207	276	345
890	6.99	14.0	21.0	27.9	34.9	41.9	48.9	55.9	62.9	69.9	105	140	175	210	279	349
900	7.07	14.1	21.2	28.3	35.3	42.4	49.5	56.5	63.6	70.7	106	141	177	212	283	353
910	7.14	14.3	21.4	28.6	35.7	42.9	50.0	57.1	64.3	71.4	107	143	179	214	286	357
920	7.22	14.4	21.7	28.9	36.1	43.3	50.6	57.8	65.0	72.2	108	144	181	217	289	361
930	7.30	14.6	21.9	29.2	36.5	43.8	51.1	58.4	65.7	73.0	110	146	183	219	292	365
940	7.38	14.8	22.1	29.5	36.9	44.3	51.7	59.0	66.4	73.8	111	148	184	221	295	369
950	7.46	14.9	22.4	29.8	37.3	44.7	52.2	59.7	67.1	74.6	112	149	186	224	298	373

For actual widths and thicknesses available, application should be made to manufacturers. Masses for greater widths and/or thicknesses than those tabulated may be obtained by appropriate addition from the range of masses given.

Mass of flats *Contd*

Kilogrammes per linear metre

Width	Thickness in millimetres															
mm	1	2	3	4	5	6	7	8	9	10	15	20	25	30	40	50
950	7.54	15.1	22.6	30.1	37.7	45.2	52.8	60.3	67.8	75.4	113	151	188	226	301	377
970	7.61	15.2	22.8	30.5	38.1	45.7	53.3	60.9	68.5	76.1	114	152	190	228	305	381
980	7.69	15.4	23.1	30.8	38.5	46.2	53.9	61.5	69.2	76.9	115	154	192	231	308	385
990	7.77	15.5	23.3	31.1	38.9	46.6	54.4	62.2	69.9	77.7	117	155	194	233	311	389
1000	7.85	15.7	23.6	31.4	39.3	47.1	55.0	62.8	70.7	78.5	118	157	196	236	314	393
1020	8.01	16.0	24.0	32.0	40.0	48.0	56.0	64.1	72.1	80.1	120	160	200	240	320	400
1040	8.16	16.3	24.5	32.7	40.8	49.0	57.1	65.3	73.5	81.6	122	163	204	245	327	408
1060	8.32	16.6	25.0	33.3	41.6	49.9	58.2	66.6	74.9	83.2	125	166	208	250	333	416
1080	8.48	17.0	25.4	33.9	42.4	50.9	59.3	67.8	76.3	84.8	127	170	212	254	339	424
1100	8.64	17.3	25.9	34.5	43.2	51.8	60.4	69.1	77.7	86.4	130	173	216	259	345	432
1120	8.79	17.6	16.4	35.2	44.0	52.8	61.5	70.3	79.1	87.9	132	176	220	264	352	440
1140	8.95	17.9	26.8	35.8	44.7	53.7	62.6	71.6	80.5	89.5	134	179	224	268	358	447
1160	9.11	18.2	27.3	36.4	45.5	54.6	63.7	72.8	82.0	91.1	137	182	228	273	364	455
1180	9.26	18.5	27.8	37.1	46.3	55.6	64.8	74.1	83.4	92.6	139	185	232	278	371	463
1200	9.42	18.8	28.3	37.7	47.1	56.5	65.9	75.4	84.8	94.2	141	188	236	283	377	471
1220	9.58	19.2	28.7	38.3	47.9	57.5	67.0	76.6	86.2	95.8	144	192	239	287	383	479
1240	9.73	19.5	29.2	38.9	48.7	58.4	68.1	77.9	87.6	97.3	146	195	243	292	389	487
1260	9.89	19.8	29.7	39.6	49.5	59.3	69.2	79.1	89.0	98.9	148	198	247	297	396	495
1280	10.0	20.1	30.1	40.2	50.2	60.3	70.3	80.4	90.4	100	151	201	251	301	402	502
1300	10.2	20.4	30.6	40.8	51.0	61.2	71.4	81.6	91.8	102	153	204	255	306	408	510
1320	10.4	20.7	31.1	41.4	51.8	62.2	72.5	82.9	93.3	104	155	207	259	311	414	518
1340	10.5	21.0	31.6	42.1	52.6	63.1	73.6	84.2	94.7	105	158	210	263	316	421	526
1360	10.7	21.4	32.0	42.7	53.4	64.1	74.7	85.4	96.1	107	160	214	267	320	427	534
1380	10.8	21.7	32.5	43.3	54.2	65.0	75.8	86.7	97.5	108	162	217	271	325	433	542
1400	11.0	22.0	33.0	44.0	55.0	65.9	76.9	87.9	98.9	110	165	220	275	330	440	550
1420	11.1	22.3	33.4	44.6	55.7	66.9	78.0	88.2	100	111	167	223	279	334	446	557
1440	11.3	22.6	33.9	45.2	56.5	67.8	79.1	90.4	102	113	170	226	283	339	452	565
1460	11.5	22.9	34.4	45.8	57.3	68.8	80.2	91.7	103	115	172	229	287	344	458	573
1480	11.6	23.2	34.9	46.5	58.1	69.7	81.3	92.9	105	116	174	232	290	349	465	581
1500	11.8	23.6	35.3	47.1	58.9	70.7	82.4	94.2	106	118	177	236	294	353	471	589
1600	12.6	25.1	37.7	50.2	62.8	75.4	87.9	100	113	126	188	251	314	377	502	628
1700	13.3	26.7	40.0	53.4	66.7	80.1	93.4	107	120	133	200	267	334	400	534	667
1800	14.1	28.3	42.4	56.6	70.7	84.8	98.9	113	127	141	212	283	353	424	565	707
1900	14.9	29.8	44.7	59.7	74.6	89.5	104	119	134	149	224	298	373	447	597	746
2000	15.7	31.4	47.1	62.8	78.5	94.2	110	126	141	157	236	314	393	471	628	785

Metric conversion of units

Measure	From metric		To metric	
	Unit	Conversion	Unit	Conversion
Length	mm	0.03937 in	in	25.4 mm (exact) — 2.54 cm
	m	3.281 ft	ft	0.3048 m — 304.8 mm
		1.094 yd	yd	0.9144 m (exact)
		0.5468 fathom		
	km	0.6214 mile	mile	1.609 km
Thickness	micron (μm)	0.03937 thou (mil or milli-inch)	thou (milli-inch)	25.4 micron or micrometre — 0.0254 mm
Area	mm²	0.00155 in²	in²	6.452 cm²
	cm²	0.1550 in²	ft²	0.0929 m²
	m²	10.76 ft² — 1.196 yd²	acre	0.4047 hectares
	hectare (100 m × 100 m)	2.471 acres	sq. miles	2.590 km²
	km²	0.3861 sq. miles		
Volume	mm³	0.00006102 in³	in³	16,390 mm³ — 16.39 cc
	m³	35.31 ft³ — 1.308 yd³	ft³	0.02832 m³
Capacity	litre	0.22 imp. gallons — 0.2542 US gallons (1 US gallons = 0.8327 imp gall)	gallon	4.546 litre
			US gallon	3.785 litre
			pint	0.568 litre (10 lb of water)
Mass	tonne (1000 kg)	0.9842 ton — 1.102 USA short ton (2000 lb)	ton (2240 lb)	1016 kg — 1.016 tonne
	kg (1000 kg)	2.205 lb	lb	0.4536 kg
	g	0.03527 oz	oz	28.35 g
Density	kg/m³	0.0624 lb/ft³	lb/ft³	16.02 kg/m³
Force	N (Newton)	0.2248 lbf — 0.1020 kgf	lbf	4.448 N — 0.4536 kgf
	kgf	2.205 lbf — 9.807 N	tonf	9.964 kN — 1.016 tonnef
	kN	0.1004 tonf — 0.2248 Kip (US) — 0.1020 tonnef	Kip (US) (1000 lbf)	4.448 kN
	tonnef (1000 kgf)	9.807 kN — 0.9842 tonf		

Metric conversion of units *Contd*

Measure	From metric — Unit	From metric — Conversion	To metric — Unit	To metric — Conversion	To metric — Conversion
Force per unit length	N/m kN/m (or N/mm) tonnef/m	0.06852 lbf/ft 0.0306 tonf/ft 0.00255 tonf/m 0.3000 tonf/ft	lbf/ft tonf/ft tonf/in	14.59 N/m 32.69 kN/m 392 kN/m	1.488 kgf/m 3.333 tonnef/m
Pressure stress or modulus of elasticity	kN/m^2 kg/cm^2 N/mm^2	0.009324 tonf/ft^2 0.01020 kgf/m^2 0.9144 tonf/ft^2 98.07 kN/m^2 145.0 lbf/in^2 0.145 ksi 10.20 kgf/cm^2 0.06475 tonf/m^2 10 bar 1 000 000 pascal	tonf/ft^2 lbf/in^2 (psi) tonf/in^2	107.3 kN/m^2 0.006895 N/mm^2 15.44 N/m^2	1.094 kg/cm^3 0.0703 kgf/cm^2 157.5 kgf/cm^2
	kgf/cm^2 atm (standard atmosphere)	14.22 lbf/m^2 0.006350 tonf/m^2 0.09807 N/N/m^2 14.70 lbf/in^2	lb/in^2	0.06805 atm	
Moment	kN/m N/m kgf/m	0.3293 tonf/ft 0.7376 lbf/ft 7.233 lbg/ft 0.1020 kgf/m 9.807 N/m	tonf/ft lbf/ft	3.037 kN/m 1.356 N/m	0.1283 kgf/m
Section Modulus (z)	cm^3	0.06102 m^3	in^3	16.39 cm^3	
Second moment of area (I)	cm^4	0.02403 m^4	in^4		41.62 cm^4
Acceleration	m/sec^2	3.281 ft/sec^2	ft/sec^2		0.3048 m/sec^2
Gravity acceleration	9.807 m/sec^2		32.17 ft/sec^2		
Velocity	km/hr m/sec	0.6214 mph 3.281 ft/sec 0.5396 UK Knots	mph ft/sec UK Knot	1.609 km/hr 0.3048 m/sec 1.853 km/hr	

Measure	From metric		To metric	
	Unit	Conversion	Unit	Conversion
Temperature	°C	$(°F - 32) \times \frac{5}{9}$	°F	$\left(°C \times \frac{9}{5}\right) + 32$
Plane angle	Radian	0.0174532 degrees $\left(\frac{\pi}{180}\right)$	degree	$57.29578 \left(\frac{180}{\pi}\right)$
Volume rate of flow	m³/sec	35.31 ft³/sec	ft³/sec (cusec)	0.02832 m³/sec
Fuel consumption	l/km	0.3540 gal/mile	gal/mile	2.825 l/km
Energy	J (Joule)	0.7376 ft/lbf	ft/lbf	1.356 J
Power	kW	1.341 hp	hp (horsepower)	745.7 W (J/sec) 0.7457 kW

SI (metric) units - multiples and submultiples

Prefix	Symbol	Factor by which the unit is multiplied		Example
tera	T	10^{12}	= 1000 000 000 000	
giga	G	10^{9}	= 1 000 000 000	gigahertz (GHz)
mega	M	10^{6}	= 1 000 000	meganewton (MN)
kilo	k	10^{3}	= 1 000	kilonewton (kN)
hecto	h	10^{2}	= 100	hectare (ha = 100 m × 100 m)
deca	da	10^{1}	= 10	
deci	d	10^{-1}	= 0.1	
centi	c	10^{-2}	= 0.01	
milli	m	10^{-3}	= 0.001	
micro	μ	10^{-6}	= 0.000 001	micrometre or micron (μm)
nano	n	10^{-9}	= 0.000 000 001	nanometre (nm)
pico	p	10^{-12}	= 0.000 000 000 001	picofarad
femto	f	10^{-16}		
atto	a	10^{-18}		

Building materials

Mass

	kN/m²	kN/m³
Aluminium roof sheeting 1.2 mm thick	0.04	
Asbestos cement sheeting		
Corrugated 6.3 mm thick as laid	0.16	
Flat 6.3 mm thick as laid	0.11	
Asphalt		
Roofing, 2 layers, 19 mm thick	0.41	
25 mm thick	0.58	
Bitumen, built up felt roofing		
3 layers including chippings	0.29	
Blockwork, excludes weight of mortar		
Concrete, solid, per 25 mm	0.54	
Concrete, hollow, per 25 mm	0.34	
Lightweight, solid, per 25 mm	0.32	
Brickwork, excludes weight of mortar		
Clay, solid, per 25 mm thick	0.45	
Low density	0.49	
Medium density	0.54	
High density	0.58	
Clay, perforated, per 25 mm thick		
Low density 25% voids	0.38	
15% voids	0.42	
Medium density 25%	0.40	
15% voids	0.46	
High density 25% voids	0.44	
15% voids	0.48	
Boards		
Cork, compressed, per 25 mm thick	0.07	
Fibre insulating, per 25 mm thick	0.07	
Laminated blockboard, per 25 mm thick	0.11	
Plywood, 12.7 mm thick	0.09	
Concrete, reinforced, 2% steel		23.55
Glass		
Clear float, 4 mm	0.09	
6 mm	0.14	

	kN/m²	kN/m³
Glass Fibre		
Thermal insulation, per 25 mm thick	0.005	
Acoustic insulation, per 25 mm thick	0.01	
Glazing, Patent		
6.3 mm Glass:		
Lead covered bars at 610 mm centres	0.29	
Aluminium alloy bars at 610 mm centres	0.19	
Lead, sheet per 3 mm thick	0.34	
Plaster		
Gypsum 12.5 mm thick	0.22	
Plasterboard Gypsum		
9.5 mm thick	0.08	
12.5 mm thick	0.11	
19.0 mm thick	0.17	
Roof Boarding		
Softwood rough sawn 19 mm thick	0.10	
Softwood rough sawn 25 mm thick	0.12	
Softwood rough sawn 32 mm thick	0.14	
Rendering		
Portland cement : sand, 1:3 mix, 12.5 mm thick	0.29	
Screeding		
Portland cement:sand, 1:3 mix, 12.5 mm thick	0.29	
Concrete, per 25 mm thick	0.58	
Lightweight, per 25 mm thick	0.32	
Steel		
Steel Roof Sheeting		77.22
0.70 mm thick (as laid)	0.07	
1.20 mm thick (as laid)	0.12	
Tiling, Roof		
Clay or concrete, plain, laid to 100 mm gauge	0.62–0.70	
Concrete, interlocking, single lamp	0.48–0.55	
Tiling, Floor		
Asphalt 3 mm thick	0.06	

Building materials

Mass

	kN/m²	kN/m³
Clay 12.5 mm thick	0.27	
Cork, compressed 6.5 mm thick	0.025	
PVC, flexible 2.0 mm thick	0.035	
Concrete 16 mm thick	0.38	
Timber		
Softwoods – Pine, Spruce,		
Douglas Fir		4.72
Redwood		5.50
Pitchpine		6.60
Hardwood – Teak, Oak		7.07
Woodwool slabs, per 25 mm thick	0.15	
Ashes, coal		7.05
Asphalt, paving		22.64
Ballast, gravel		19.20
Brick		20.00
Cement, Portland loose		14.11
Cement, mortar		16.46
Clay, damp, plastic		17.54
Concrete, breeze		15.09
Concrete, brick		18.82
Concrete, stone		22.64
Earth, dry, loose		11.30
Earth, moist, packed		15.09
Earth, dry, rammed		17.54
Glass, plate		27.34
Glass, sheet		24.50
Gravel		18.82
Lime mortar		16.17
Masonry, artificial stone		22.60
Masonry, freestone, dressed		25.00
Masonry, freestone, rubble		21.95
Masonry, granite, dressed		31.00
Masonry, granite, rubble		24.30

	kN/m³
Aluminium, cast	27.50
Brass, cast	87.00
Brass, rolled	83.84
Bronze	82.27
Copper, cast	90.00
Copper, rolled	87.60
Iron, cast	72.00
Iron, wrought	76.80
Lead, cast	111.13
Lead, sheet	111.42
Nickel, monel metal	89.00
Steel, cast	77.22
Steel, rolled	77.22
Tin, cast	72.80
Tin, rolled	72.52
Zinc	68.60
Pitch	11.50
Plaster	15.09
Plaster of Paris, set	12.54
Sand, dry	16.00
Sand, wet	20.00
Slate	29.00
Flint	25.90
Granite	31.00
Limestone	28.00
Macadam	23.57
Marble	28.00
Sandstone	25.00
Tar	12.00
Terra-cotta	17.90

Packaged materials

Mass

	kN/m³		kN/m³
Cereals etc.		Oils, in barrels	5.65
Barley, in bags	5.65	Oils, in drums	7.07
Barley, in bulk	6.28	Paper, printing	6.28
Flour, in bags	7.07	Paper, writing	9.42
Hay, in bales, compressed	3.77	Petrol	6.59
Hay, not compressed	2.20	Plaster, in barrels	8.32
Oats, in bags	4.24	Potash	32.14
Oats in bulk	5.02	Red lead, dry	20.72
Potatoes, piled	7.07	Rosin, in barrels	7.54
Straw, in bales compressed	2.98	Rubber	9.42
Wheat, in bags	6.12	Saltpetre	10.52
Wheat, in bulk	8.50	Screw nails, in packages	15.70
		Soda ash, in barrels	9.73
Miscellaneous		Soda, caustic, in drums	13.82
Bleach, in barrels	5.02	Snow, freshly fallen	0.94
Cement, in bags	13.19	Snow, wet, compact	3.14
Cement, in barrels	11.46	Starch, in barrels	3.93
Clay, china, kaolin	21.67	Sulphuric acid	9.42
Clay, potters, dry	18.84	Tin, sheet, in boxes	43.65
Coal, loose	8.79	Water, fresh	9.81
Coke, loose	4.71	Water, sea	10.05
Crockery, in crates	6.28	White lead, dry	13.50
Glass, in crates	9.42	White lead paste, in drums	27.32
Glycerine, in cases	8.16	Wire, in coils	11.62
Ironmongery, in packages	8.79		
Leather, in bundles	2.51		
Leather, hides, compressed	3.61		
Lime, in barrels	7.85		
Oils, in bulk	8.79		

Angle of internal friction and mass of materials

Material	Mass in kN/m³	Angle of internal friction°
Ashes	6.3 – 11.6	20 – 40°
Cement	13.4 – 16.8	20°
Cement clinker	14.0 – 16.0	30 – 35°
Chalk (in lumps)	11.0 – 22.0	35° – 45°
Clay		
in lumps	11.0	30°
dry	18.8 – 22.0	30°
moist	20.4 – 25.1	45°
wet	20.4 – 25.1	15°
Clinker	10.0 – 15.0	30 – 40°
Coal (in lumps)	8.0 – 19.0	20 – 45°
Coke	4.0 – 6.0	30°
Copper ore	25.1 – 29.2	35°
Crushed brick	12.6 – 21.8	35° – 40°
Crushed stone	17.3 – 20.4	35° – 40°
Granite	17.3 – 31.0	35° – 40°
Gravel (clean)	14.1 – 20.0	35° – 40°
Gravel (with sand)	15.7 – 19.2	25° – 30°
Haematite iron ore	36.1	35°
Lead ore	50.0 – 52.0	35°
Limestones	12.6 – 18.8	35° – 45°
Magnetite iron ore	40.0	35°
Manganese ore	25.1 – 28.8	35°
Mud	16.5 – 22.8	0°
Rubblestone	17.3 – 19.8	45°
Salt	7.7 – 9.6	30°
Sand		
dry	15.7 – 18.8	30° – 35°
moist	18.1 – 19.6	35°
wet	18.1 – 20.4	25° – 30°
Sandstones	12.6 – 25.0	35° – 45°
Shale	14.1 – 19.8	30° – 35°
Shingle	14.1 – 17.3	30° – 35°
Slag	14.1 – 24.8	35°

Material	Mass in kN/m³	Angle of internal friction°
Vegetable earth		
dry	14.1 – 15.7	30°
moist	15.7 – 17.3	45° – 50°
wet	17.3 – 18.8	15°
Zinc ore	25.1 – 28.3	35°

All materials should be tested under appropriate conditions prior to use in final design.

Values of K_a (coefficient of active pressure) for cohesionless materials

This table may be used to determine the horizontal pressure exerted by material, p_a, in kN/m².

p_a = mass × depth of material × K_a

Values of δ	Values of K_a for values of Ø				
	25°	30°	35°	40°	45°
0°	0.41	0.33	0.27	0.22	0.17
10°	0.37	0.31	0.25	0.20	0.16
20°	0.34	0.28	0.23	0.19	0.15
30°	–	0.26	0.21	0.17	0.14

The effect of wall friction δ on active pressures is small and is usually ignored. The above values of K_a assume vertical walls with horizontal ground surface.

Note: The above data should *not* be used in the design calculations for silos, bins, bunkers and hoppers.

Approximate mass of floors

Reinforced concrete floors

	Mass in kN/m²	
Thickness	Dense concrete	Lightweight concrete
100	2.35	1.76
125	2.94	2.20
150	3.53	2.64
175	4.11	3.08
200	4.70	3.52
225	5.23	3.96
250	5.88	4.40

Dense concrete is assumed to have natural aggregates and 2% reinforcement with a mass of 2400 kg/m³.

Lightweight concrete is assumed to have a mass of 1800 kg/m³.

Steel floors

Durbar non-slip		Open steel flooring		
Thickness on plain mm	Mass in kN/m²	Thickness mm	Mass in kN/m² Light	Heavy
4.5	0.37	20	0.29	0.38
6.0	0.49	25	0.38	0.46
8.0	0.64	30	0.44	0.56
10.0	0.80	40	0.60	0.74
12.5	0.99	50	0.74	0.90

Open steel floors are available from various manufacturers to particular patterns and strengths.

The above average figures are for guidance in preliminary design. Manufacturers' data should always be used for final design.

Timber floors

Solid timber, joist sizes, mm. Mass in kN/m²

Joist Centres	Decking	75 × 50	100 × 50	150 × 50	200 × 50	225 × 50	275 × 50
400 mm	19 mm Softwood	0.16	0.18	0.21	0.25	0.27	0.30
	19 mm Chipboard	0.19	0.21	0.24	0.28	0.30	0.33
	22 mm Chipboard	0.21	0.23	0.26	0.30	0.32	0.35
600 mm	19 mm Softwood	0.14	0.16	0.18	0.20	0.21	0.24
	19 mm Chipboard	0.17	0.19	0.21	0.23	0.24	0.27
	22 mm Chipboard	0.19	0.21	0.23	0.25	0.26	0.29

The solid timber joists are based on a density of 5.5 kN/m³.

Walls and partitions - mass

Walls

Construction	kN/m²		
	Brick	Block	Brick + Block
102.5 mm thick			
Plain	2.17	1.37	
Plastered one side	2.39	1.59	
Plastered both sides	2.61	1.81	
215 mm thick			
Plain	4.59	2.99	3.79
Plastered one side	4.81	3.21	4.01
Plastered both sides	5.03	3.43	4.23
255 mm Cavity wall			
Plain	4.34	2.74	3.54
Plastered one side	4.56	2.96	3.76
Plastered both sides	4.78	3.18	3.98

Assumed mass of brickwork 21.2 kN/m³

Assumed mass of blockwork 13.3 kN/m³

Walls and partitions - mass

Partitions

	kN/m²
Timber partition (12.5 mm plasterboard each side)	0.25
Studding with lath and plaster	0.76

For specific types and makes of walls and partitions, reference should be made to the manufacturers' publications.

Areas and volumes

Areas

Parallelogram	= base × perpendicular height
Triangle	= base × ½ perpendicular height
Trapezoid	= ½ sum of parallel sides × perpendicular height
Circle	= .7854 × square of diameter
Sector of circle	= length of arc × ½ radius
Parabola	= base × ⅔ height
Ellipse	= long diameter × short diameter × .7854
Regular polygon	= sum of sides × ½ perpendicular distance from centre to sides
Surface of sphere	= π × square of diameter
Surface of cone	= area of base + (circumference of base × ½ slant height)

Volumes

Prism	= area of base × height
Pyramid or cone	= area of base × ⅓ height
Sphere	= 4.1888 × radius²

Positions of centre of gravity

Triangle	= ⅓ perpendicular height from base
Parabola	= ⅔ height from base
Pyramid or cone	= ¼ height from base

Side of square of equal area to circle = diameter × .8862
Diameter of circle of equal area to square = side × 1.1284
Circumference of circle = π × diameter

Metric equivalents of standard wire gauges

Standard wire gauge	Dia mm	Standard wire gauge	Dia mm	Standard wire gauge	Dia mm
4/0	10.16	3	6.40	9	3.66
3/0	9.45	4	5.89	10	3.25
2/0	8.84	5	5.39	11	2.95
1/0	8.23	6	4.88	12	2.64
1	7.62	7	4.47	13	2.34
2	7.01	8	4.06	14	2.03

The Greek alphabet

Name	Capital Letter	Small Letter	English Equivalent	Name	Capital Letter	Small Letter	English Equivalent
Alpha	A	α	a	Nu	N	ν	n
Beta	B	β	b	Xi	Ξ	ξ	x
Gamma	Γ	γ	g	Omicron	O	ο	short 0
Delta	Δ	δ	d	Pi	Π	π	p
Epsilon	E	ε	short e	Rho	P	ρ	rh
Zeta	Z	ζ	z	Sigma	Σ	σ	s
Eta	H	η	long e	Tau	T	τ	t
Theta	Θ	θ	th	Upsilon	Y	υ	u
Iota	I	ι	i	Phi	Φ	φ	ph
Kappa	K	κ	k	Chi	X	χ	ch
Lambda	Λ	λ	l	Psi	Ψ	ψ	ps
Mu	M	μ	m	Omega	Ω	ω	long 0

Circular arcs

The following formulae may be used for exact geometrical calculations.

For	Expressions			
N ratio	$\dfrac{R}{2L}$	$\dfrac{1}{\theta} - \dfrac{\theta}{12} - \dfrac{\theta^3}{720} - \dfrac{\theta^5}{30240}$ etc. (θ Radians)		$\sqrt{\dfrac{R^2}{C^2} - \dfrac{1}{4}}$
θ length	$\text{Sin}\,\theta = \dfrac{T}{R}$; $\;R \times \theta$ radians	$\sqrt{\dfrac{R}{2Q} - \dfrac{1}{4}}$; $\;\cos\theta = 1 - \dfrac{C^2}{2R^2}$	$\dfrac{T}{2Q}$; $\;\cos\theta = \sqrt{\dfrac{R^2 - T^2}{R}}$	$\dfrac{R \pm \sqrt{R^2 - T^2}}{2T}$; $\;\tan\theta = \dfrac{T}{\sqrt{R^2 - T^2}}$
C chord length	$\dfrac{R}{\sqrt{N^2 + \frac{1}{4}}}$	$\sqrt{2RQ}$	$\sqrt{T^2 + Q^2}$	$\sqrt{E^2 - Q^2}$
T	$\dfrac{RN}{N^2 + \frac{1}{2}}$	$2QN$	$\sqrt{2RQ - Q^2}$	
Q	$\dfrac{R}{2(N^2 + \frac{1}{4})}$	$\dfrac{C^2}{2R}$	$R - \sqrt{R^2 - T^2}$	$\dfrac{T}{2N}$
R radius	$\sqrt{C^2 - T^2}$; $\;C\sqrt{N^2 + \tfrac{1}{2}}$	$2Q\left(N^2 + \tfrac{1}{2}\right)$	$2LN$	$\dfrac{NW}{\sqrt{N^2 + \frac{1}{4}} - N}$; $\;\dfrac{C^2}{8V} + \dfrac{V}{2}$
V versine	$R - \tfrac{1}{2}\sqrt{4R^2 - C^2}$	$R - CN$	$R\left(1 - \dfrac{N}{\sqrt{N^2 + \frac{1}{4}}}\right)$	$\dfrac{T\left(N^2 + \frac{1}{4}\right)}{N}\left(1 - \dfrac{N}{\sqrt{N^2 + \frac{1}{4}}}\right)$
L	$\dfrac{R}{2N}$	$\dfrac{T^2 + Q^2}{2Q}$		
W	$R\left(\dfrac{\sqrt{N^2 + \frac{1}{4}} - 1}{N}\right)$	$\dfrac{T}{2} + \dfrac{Q^2}{2T}$		
W + V	$\dfrac{L}{2\sqrt{N^2 + \frac{1}{4}}}$	$2(W + V)$		
A	$\dfrac{QC}{T}$			
B	$\dfrac{C^2}{T}$			
Y	$Y - R + \sqrt{R^2 - X}$			

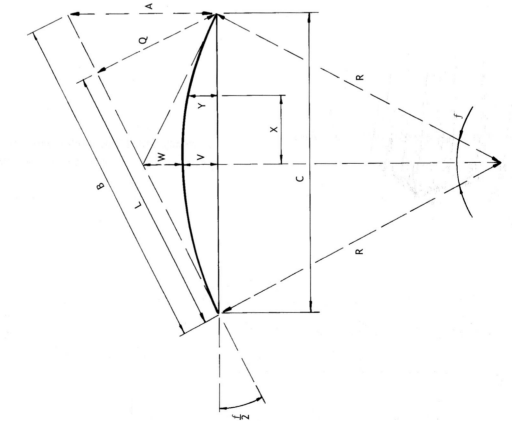

Worked example

Question

A beam is 20 m long and is to be cambered to a circular vertical curve of radius 60 m.

Find

(a) vertical offset at mid-length
(b) vertical offset at $^1/_4$ points
(c) slope of beam at ends
(d) true length of beam

Answer

(a) offset at mid length (or versine)

$$v = R - \frac{1}{2}\sqrt{4R^2 - C^2}$$

$$= 60 - \frac{1}{2}\sqrt{4 \times 60^2 - 20^2}$$

$$= 0.839\,m$$

(b) At $\frac{1}{4}$ point

$$X = \frac{C}{4} = \frac{20}{4} = 5.000\,m$$

$$y = v - R + \sqrt{R^2 - X^2}$$

$$= 0.839 - 60 + \sqrt{60^2 - 5^2}$$

$$= 0.630\,m$$

(c) Slope of beam at ends

$$ws\ \theta = 1 - \frac{C^2}{2R^2} = 1\frac{20^2}{2 \times 60^2} = 0.94444$$

$$\therefore\ \theta = 19.188° \text{ or } 0.3349 \text{ radians}$$

$$\text{Slope at ends} = \frac{\theta}{2} = 9.594° \text{ or } 0.1675 \text{ radians}$$

(d) Arc length $= R \times \theta$ radius
$$= 60 \times 0.3449 \text{ radians}$$
$$= 20.094\,m$$

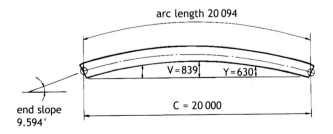

arc length 20 094

V=839   Y=630

C = 20 000

end slope 9.594°

Circular arcs – large radius to chord ratios

The following simplified formulae are approximate but are usually sufficiently accurate, typically when

$$\frac{R}{C} > 5 \text{ or } \frac{C}{V} > 40 \text{ or } \frac{X}{Y} > 20$$

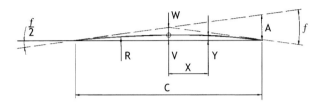

$$V = \frac{C^2}{8R} = W \qquad R = \frac{C^2}{8V}$$

$$A = 4V = \frac{C^2}{2R} \qquad \frac{\theta}{2} = \frac{C}{2R} = \frac{4V}{C}$$

$$Y = V\left(1 - \frac{4X^2}{C^2}\right)$$

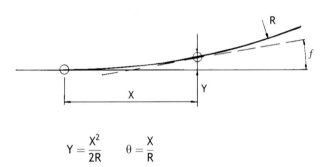

$$Y = \frac{X^2}{2R} \qquad \theta = \frac{X}{R}$$

Precamber for a simply Supported Beam

The following formulae can be used to provide deflection and slope values for a beam of uniform stiffness which is uniformly loaded. This enables a precise precamber shape to be determined so as to counteract deflection. The shape will generally be suitable for beams which are not loaded uniformly. Often a circular or parabolic profile is adopted in practice, and is sufficiently accurate.

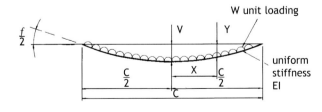

W unit loading

uniform stiffness EI

Deflected form

Central deflection:

$$V = \frac{5}{384}\frac{WC^4}{EI}$$

Rotation at ends:

$$\frac{\theta}{2} = \frac{WC^3}{24EI}$$

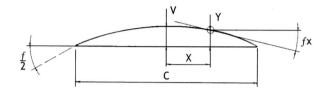

Precambered form to counteract deflection

Precamber at any point:

$$Y = V\left(1 - 4.8\left(\frac{X}{C}\right)^2 + 3.2\left(\frac{X}{C}\right)^4\right)$$

Slope at any point:

$$\theta x = \frac{\theta}{2}\left(\frac{3X}{C} - 4\left(\frac{X}{C}\right)^3\right)$$

Parabolic arcs

The following formulae may be used for calculations of parabolic arcs which are often used for precambering of beams.

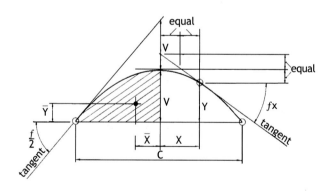

$$\frac{\theta}{2} = \frac{4V}{C}$$

$$Y = V\left(1 - \frac{4X^2}{C^2}\right)$$

$$\theta x = \frac{8VX}{C^2}$$

Approximate arc length =

$$2\sqrt{\left(\frac{C}{2}\right)^2 + \frac{4}{3}V^2} \quad \text{where} \quad \frac{V}{C} < 0.05$$

For shaded area under curve:

$$\text{Area} = \frac{2}{3} \times \left(\frac{C}{2} \times V\right)$$

$$\bar{X} = 0.375 \times \left(\frac{C}{2}\right)$$

$$\bar{Y} = 0.4 \times V$$

Braced frame geometry

Given	To find	Formula
bpw	f	$\sqrt{(b+p)^2 + w^2}$
bw	m	$\sqrt{b^2 + w^2}$
bp	d	$b^2 \div (2b + p)$
bp	e	$b(b + p) \div (2b + p)$
bfp	a	$bf \div (2b + p)$
bmp	c	$bm \div (2b + p)$
bpw	h	$bw \div (2b + p)$
afw	h	$aw \div f$
cmw	h	$cw \div m$

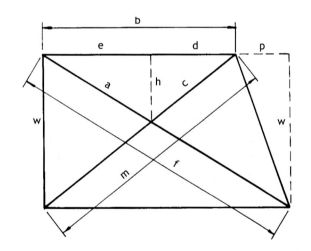

Given	To find	Formula
bpw	f	$\sqrt{(b+p)^2 + w^2}$
bnw	m	$\sqrt{(b-n)^2 + w^2}$
bnp	d	$b(b - n) \div (2b + p - n)$
bnp	e	$b(b + p) \div (2b + p - n)$
bfnp	a	$bf \div (2b + p - n)$
bmnp	c	$bm \div (2b + p - n)$
bnpw	h	$bw \div (2b + p - n)$
afw	h	$aw \div f$
cmw	h	$cw \div m$

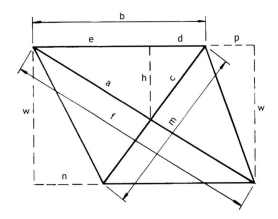

Given	To find	Formula
bpw	f	$\sqrt{(b+p)^2 + w^2}$
bkv	m	$\sqrt{(b+k)^2 + v^2}$
bkpvw	d	$bw(b + k) \div [v(b + p) + w(b + k)]$
bkpvw	e	$bv(b + p) \div [v(b + p) + w(b + k)]$
bfkpvw	a	$fbv \div [v(b + p) + (w(b + k)]$
bkmpvw	c	$bmw \div [v(b + p) + w(b + k)]$
bkpvw	h	$bvw \div [v(b + p) + w(b + k)]$
afw	h	$aw \div f$
cmv	h	$cw \div m$

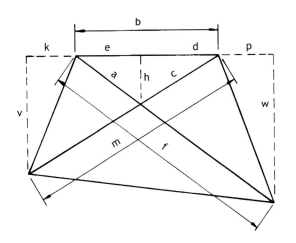

Parallel bracing

k = (log B − log T) ÷ no. of panels. Constant k plus the logarithm of any line equals the log of the corresponding line in the next panel below.

$$a = TH \div (T + e + p)$$

$$b = Th \div (T + e + p)$$

$$c = \sqrt{\left(\tfrac{1}{2}T + \tfrac{1}{2}e\right)^2 + a^2}$$

$$d = ce \div (T + e)$$

$$\log e = k + \log T$$

$$\log f = k + \log a$$

$$\log g = k + \log b$$

$$\log m = k + \log c$$

$$\log n = k + \log d$$

$$\log p = k + \log e$$

The above method can be used for any number of panels. In the formulas for 'a' and 'b' the sum in parenthesis, which in the case shown is (T + e + p), is always composed of all the horizontal distances except the base.

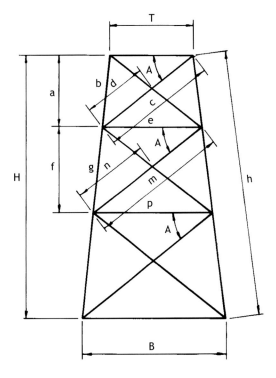

Index

abbreviations, table 2.2, 40

angle section

   backmarks, 63

   sizes, table 4.7, 61

areas, 161

bars, mass of round & square sections, 148

beam splice details, figure 5.4, 87

beams, universal sizes, table 4.3, 55

bolt edge distances, 72

bolt load capacities, 49

bolted connections, load capacity,

      table 3.1, 43

bolted trusses, figure 5.7, 90

bolting, 20

bolts

  black

    dimensions of, 53

    mechanical properties, table 4.1, 53

  details, 52

  HSFG

    coronet load indicator, figure 1.26, 22

    dimensions of, table 4.2, 54

    mechanical properties, table 4.2, 54

  types used in UK, table 1.10, 21

braced frame geometry formula, 164

bracing details, figure 5.5, 88

bridges, 121

  abutment detail, figure 7.28, 131

  cross sections

    beams, figure 7.21, 123

    box girders, figure 7.22, 124

    railway, figure 7.22, 124

  parapet rail details, figure 7.24, 126

  precamber sketch, figure 7.25, 129

  splice detail, figure 7.29, 132

building materials, self weights, 156

buildings, 34

bulb flat cross sections, 68

bulb flats sizes, table 4.11, 68

butt weld shrinkage, 14

cad modelling, 96

  figure 6.1, 97

cad connection library, figure 6.2, 98

cnc/rapid prototyping, figure 6.3, 100

camber distortion, 12

centre of gravity, 161

channel section sizes, table 4.6, 60

circular arcs

  properties, 162

  worked example, 163

clearances around highways and railways,

      figure 4.2, 77

cold formed sections, 7

column base

  details, 48

  details of holding down bolts, 48

  size and load capacity, 42

column splice detail, figure 5.3, 86

columns, universal sizes, table 4.4, 58

composite railway bridge, figure 7.22, 124

composite construction, figure 1.1, 2

  in building floors, 104

computer aided detailing, 95

computer draughting systems, 95

connection, load capacities, 41

  typical details, 84

connections, 15

  beam to beam continuous, figure 5.2, 85

  beam to column continuous, figure 5.1, 84

  beam to column simple, figure 5.1, 84

  beam to column with angle cleats, 44

  beam to column with end plates, 44

  bracing details, figure 5.5, 88

  dos and don'ts, figure 1.27, 24, 25

Steel Detailers' Manual, Third Edition. Alan Hayward and Frank Weare. Third edition revised by Anthony Oakhill.
© 2011 Alan Hayward, Frank Weare and Anthony Oakhill. Published 2011 by Blackwell Publishing Ltd.

connections (*Continued*)

    hollow sections, figure 5.6, 89

    hot rolled and hollow tubes, figure 1.18, 17

    moment/rotation, figure 1.15, 15

    precast concrete floors, figure 5.11, 94

    simple, table 3.1, 43

    site locations, figure 1.17, 16

    simple and continuous, figure 1.16, 16

    steel to timber, figure 5.10, 93

    worked examples, 41

conversion of units, 153

corrosion, dos and don'ts, figure 1.29, 28

crane gantry girder details, figure 7.8, 107

crane rail cross section, 69

crane rails, table 4.12, 69

curved sections (about major axis),

       table 1.7, 7

design guidance, 41

detailing

    abbreviations, 40

    bolts, 39

    conventions, 37

    data, 52

    opposite handing, 39

    welds, 39

dimensional variations, table 1.8, 12

dimensions, 36

drawing

    layout, figure 2.1, 36, 37

    marking system, figure 2.2, 38

    projection, 37

    revisions, 37

    scales, 37

drawings, 30

durbar floor plate, table 4.14, 73

electric overhead travelling cranes, 107

engineer's drawings, 30

environmental conditions, effect on steel, 32

erection

    lifting beam, 134

    marks, 38

fabrication sequence, 18

fire protection, board connection, figure 7.6, 105

fire resistance, 105

flats, mass of, 149

floor plates, 73

floors, self weights, 160

footbridges

    cross sections, figure 7.23, 125

    handrail details, 125

foundations, interface with structure, 16

gable ends – portal frame building, figure 7.9, 108

galvanizing, 30

girders

    gantry girders, 107

    lattice girders, 91

grit blasting, 30

headroom requirements, 77

highway

    clearances, figure 4.2, 77

    sign gantry, 135

holding down bolts, 39

hollow sections

    circular sizes, table 4.10, 66

    rectangular sizes, table 4.9, 65

    square sizes, table 4.8, 64

    weld preparations, figure 4.5, 80

HSFG bolts, 22

    dimensions, 54

    load capacity, 54

HSFG power wrench details, 71

joist section sizes, table 4.5, 58

ladders, layout, figure 4.1, 75

lamellar tearing, 19

    details, figure 1.24, 20

lattice girders, figure 5.8, 91

lettering on drawings, 36

lifting beam, figure 7.30, 134

load factors and combinations, 34

marking system, figure 2.1, 37

materials

    self properties, 159

    self weights, 156

metal coatings, 30

multi-storey frame buildings, 102

natural light requirements, 106

omnia plank, connection to beams, figure 5.11, 94

packaged materials self weight, 158

paint treatments, 31

plate girder
    cross section details, figure 7.26, 129
    splice detail, figure 7.29, 132
plates, available lengths
    normalised condition, table 4.15, 82
    normalised rolled condition, table 4.16, 83
portal frame building
    details, 110
    gable ends, figure 7.10, 110
    roof bracing, figure 7.10, 110
portal frame buildings, 107
precamber for simply supported beam, 163
projection – third angle, 37
protective treatment, 24
    systems, table 1.13, 31
purlins – cold rolled sections, 109

railway clearance, figure 4.2, 77
references, 143
rolled sections, tolerances and effects,
        figure 1.8, 13
roof over reservoir, 114

single storey building, cross sections, figure 7.7, 106
single storey buildings, 106
site bolting, 16
site welding, 16
splice detail, bridge plate girder, 132
splices
    in beams, figure 5.4, 87
    in columns, figure 5.3, 86
    in hollow sections, figure 5.6, 89
staircase, 139
stairs, layout, figure 4.1, 75
steel
    advantages of its use, 1
    guidance on grades, table 1.4, 5
    main use of steel grades, table 1.3, 4
    properties, table 1.2, 4
    recommended grades, 3
    requirements, 2
    stress strain curve, 3
    weather resistant types, 4
step ladders, 76
structural shapes, figure 1.6, 6, 9
structural tolerances, 8
structures
    bridges, 121
    highway sign gantry, 135

multi-storey frame buildings, 102
portal frame buildings, 107
roof over reservoir, 115
single span highway bridge, 128
single-storey frame buildings, 106
staircase, 139
tower, 117
vessel support structure, 110
symbols, the Greek alphabet, 161

timber, connection to steelwork, 93
tolerances, 8
    rolled sections, figure 1.8, 13
tower structure, 117
transport sizes, maximum, figure 4.3, 78
truss details, bolted and welded, figure 5.7, 90
tube sizes, see hollow sections, 64
twisting of angles and channels, figure 1.5, 8

units, conversion table, 153
universal beam, hole spacing, 55
universal column
    hole spacing, 58
    sizes, 58

vessel support structure, 110
volumes, 161

walkways, see stairs and ladders, 75
walls and partitions, self weights, 160
weld
    load capacities, table 3.7, 51
    preparation details, figure 4.5, 80
    process, table 1.9, 19
    shrinkage, worked example, 12
    size considerations, 18
    symbols, figure 4.4, 79
    type choice, 19
    types, fillet and butt welds, 18
welded trusses, figure 5.7, 90
welding, 16
welding distortion, 8
    worked example for plate girder, 12
welding distortions, figure 1.7, 11
wire gauges, standard, 161
workshop drawing
    beam detail, figure 7.4, 104
    column detail, figure 7.5, 105
workshop drawings, 33